DOWNHOLE PRODUCTION AND SUBSEA PROCESSING

This volume consists of papers presented at the Downhole Production and Subsea Processing Conference, held in Aberdeen, UK on 3 November 1998.

ACKNOWLEDGEMENTS

The valuable assistance of the Technical Advisory Committee and panel of referees is gratefully acknowledged.

TECHNICAL ADVISORY COMMITTEE

Dr Alan Burns	Elf UK
Ms Annette Cutler	Technology Initiatives Limited
Dr Derek Mathieson	Shell UK Exploration and Production
Dr M M Sarshar	BHR Group Limited
Mr Andy Thomson	Halliburton Manufacturing and Services Limited
Mr Richard Winchester	The Centre for Marine and Petroleum Technology

CO-SPONSOR

The Centre for Marine and Petroleum Technology – CMPT®

ORGANIZED AND SPONSORED BY

BHR Group Limited
Cranfield
Bedfordshire MK43 0AJ
Tel: +44 (0) 1234 750422
Fax: +44 (0) 1234 750074
EMail: ccox@bhrgroup.co.uk
World Wide Web: http://www.bhrgroup.co.uk/confsite/index.html

CONFERENCE ON

DOWNHOLE PRODUCTION AND SUBSEA PROCESSING

TECHNICAL, BUSINESS, AND ENVIRONMENTAL SOLUTIONS FOR THE INDUSTRY

Edited by J A R Muir

BHR Group Conference Series
Publication No.34

Papers presented at the *Downhole Production and Subsea Processing Conference*, organized and sponsored by BHR Group Limited.
Held in Aberdeen, UK, on 3 November 1998.

Professional Engineering Publishing Limited, Bury St Edmunds and London, UK.

Excellence in Fluid Engineering

ISBN 1 86058 176 5

Printed and bound in Great Britain by Bookcraft (Bath) Limited.

A CIP catalogue record for this book is available from the British Library.

Other Titles in the BHR Group Conference Series:

Related Titles of Interest

Title	Author/Editor	ISBN
Plant Monitoring and Maintenance Routines	IMechE Seminar 1998–2	1 86058 087 4
Fluid Machinery for the Oil, Petrochemical, and Related Industries	IMechE Conference 1996–4	0 85298 994 6
Subsea Control and Data Acquisition	Edited by L Adriaansen, R Phillips, C Rees, and J Cattanach	0 85298 993 8

For the full range of titles published by Professional Engineering Publishing contact:

Sales Department
Professional Engineering Publishing Limited
Northgate Avenue
Bury St Edmunds
Suffolk
IP32 6BW
UK
Tel: +44 (0) 1284 724384
Fax: +44 (0) 1284 718692

Contents

Slimwell construction 'without the pain'

P HEAD
XL Technology Limited
G CAMERON
Amerada Hess
T HANSON
Enterprise Oil
S AL-RAWAHI
Shell International, UK

SYNOPSIS

Slimming down the well geometry to reduce the cost of well construction has been an economic goal of operators for many years[1,2,3,4,5,6,7] Historically. this has resulted in small diameter holes across the reservoir with the associated problems of specialised equipment requirements. reduced drilling rates. complicated well evaluation and reduced flow path. In addition. systems to date have also introduced severe operational drawbacks such as high surge and swabbing pressures during casing installation and high circulation pressures while cementing.

This paper will present the results from a joint industry project (JIP) the object of which was to address these issues. and provide an effective solution to overcome these drawbacks. The common approach throughout all of the well construction processes described in this paper is a series of liners. with tiebacks to previous liner tops as required. Finally the well is completed with a monobore tieback to surface.

The main technical challenges to overcome were:

1. How to terminate each liner hanger with so little annular space while still retaining the same tensile load capacity as the virgin pipe. This was achieved by engineering a unique liner hanger system which requires no moving components, and which forms a metal to metal seal with the previous liner.

2. To prevent the open hole being exposed to excessive pressures during liner installation, and while cementing. This was resolved by a novel conveyance system which allows for two modes of fluid circulation while the liner is being deployed and landed in the well, with a third conventional mode for circulating and cementing operations.

With the current trend towards smaller riser and casing configurations. particularly in deep water operations[8]. this innovative development now allows a broader range of options for well architects. Unlike current practice, wells may now be designed to fit within this reducing riser envelope while maintaining conventional size completions across the reservoir.

INTRODUCTION

Development Philosophy
The development process for this project has been to formulate a method of overcoming the traditional drawbacks to micro annular clearances. The project includes, the design, manufacture and bench testing of the individual tools, with a full scale test planned to demonstrate the operation of the entire system.

Phase I
The first phase of the project is to use flush jointed pipe. Apart from some issues regarding handling the pipe, this provides a quick and straight forward development path for the various tools and components.

Phase II/III
Further phases of the project aim to slim the clearance much further and this will be achieved by using either amorphous butt welded joints of pipe or seam welded coiled tubing which is currently commercially available in sizes up to 6.5/8" diameter.

The latter two are driven more by the availability of equipment to handle and accommodate large diameter coiled tubing. While the technologies are not mature, Amorphous welding holds great promise but as yet it is only performed in a controlled environment. However, as can be seem in figure 1, the welded options offer the smallest clearances and it is anticipated that this is where the largest savings will be gained.

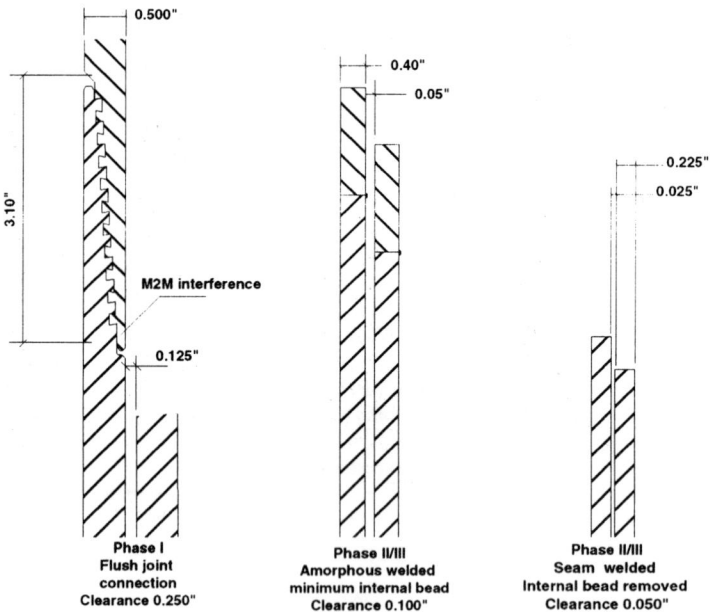

Figure 1. Microannulus casing options.

WELLBORE ARCHITECTURE

Figure 2. shows the various wellbore geometries which can be achieved using this technology. Three casing options are compared to a traditional well design. A comparison is made with the amount of materials to be removed in the form of drill cuttings, and the materials (casing and cement) which are then required to provide structural support and hydraulically isolate the conduit to the reservoir.

The reduction in size also has an impact on the drilling fluid selection. Smaller hole volumes make the selection of high quality drilling fluid more economic. This in turn enhances the drilling process, reduces the risk of other-hole / drilling fluid related problems and finally assists in the mechanical location and operation of the conveyance and landing hardware.

The coiled casing system offers additional benefits when combined with coiled tubing drilling, as it could potentially allow the casing to be deployed with the well at balance. In addition the cementing operation could be performed similarly at balance, significantly reducing the risk of lost circulation, formation damage and gas migration.

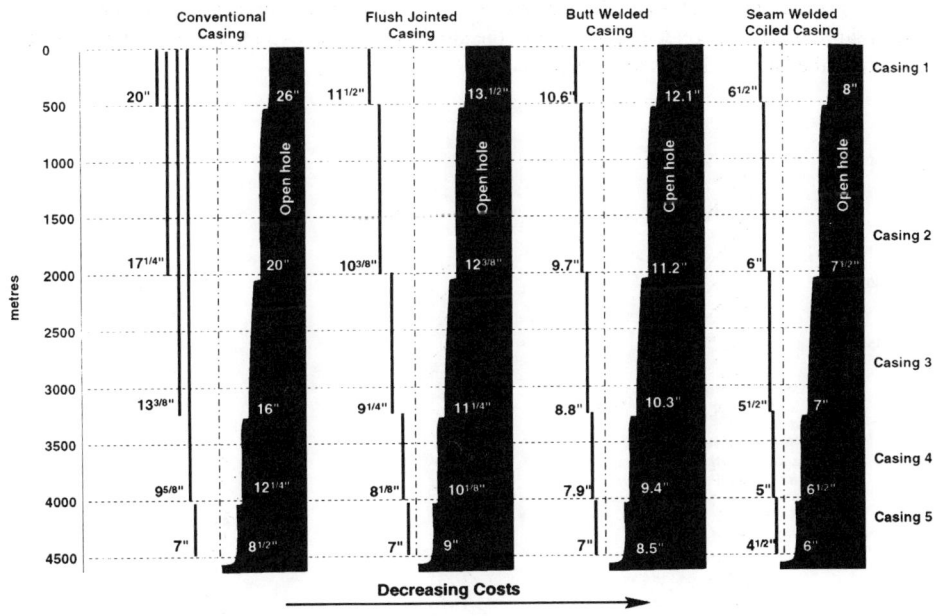

Figure 2. Casing options. for 7" casing to Reservoir and CT Casing system alternative

Increased Contingency Strings

Additional casing strings may be accommodated within the same dimensional boundaries as conventionally constructed wells. Figure 3 has been included to show how this can impact on options for additional contingency strings without severely impacting the size of the borehole across the reservoir. In one example the use of flush jointed pipe has been highlighted, and the second example illustrates a construction process using a combination of flush jointed pipe and seam welded pipe. This could offer advantages in salt stringer sections, or in HP/HT wells where additional setting depths are sometimes unavoidable.

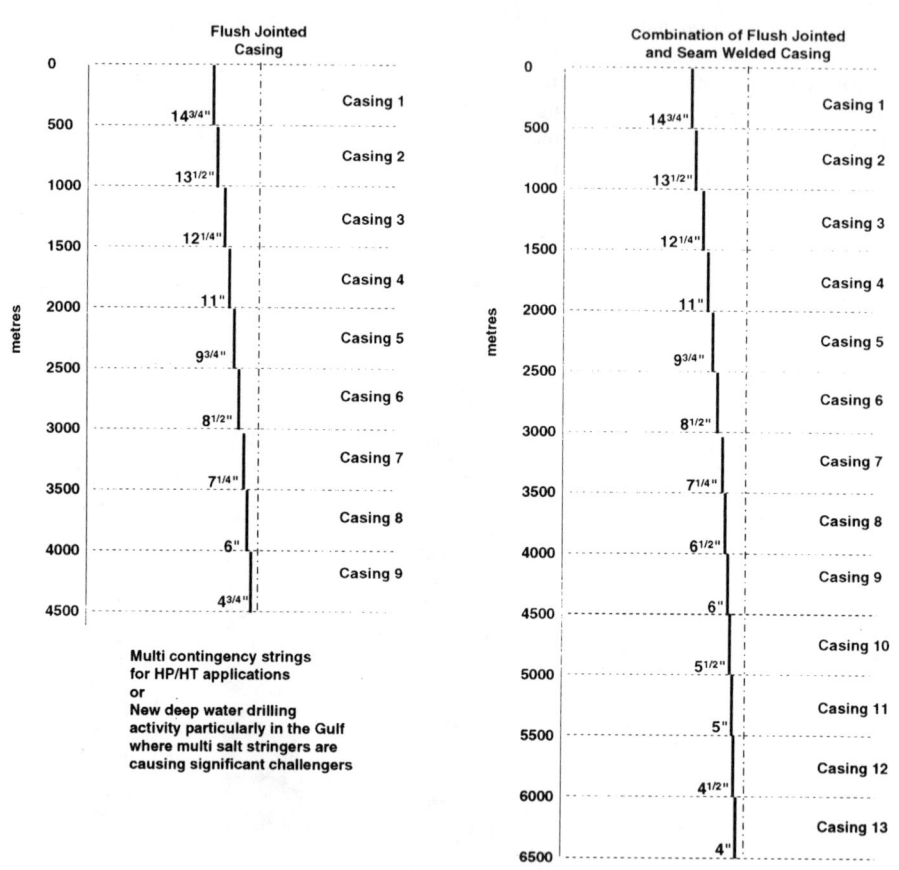

Figure 3. Increased Contingency Options

Alternate Liner / Completion tieback options through the drilling process

Tie back liners can be installed as the well is deepened, to provide "sacrificial" wear liners for the drilling process, provide additional pressure containment barriers or increase the structural integrity across a particular high pressure zone. Finally, the monobore production tie back string could be installed under live well conditions if large diameter coiled tubing was used. Similarly it could be retrieved under live conditions on a predetermined basis.

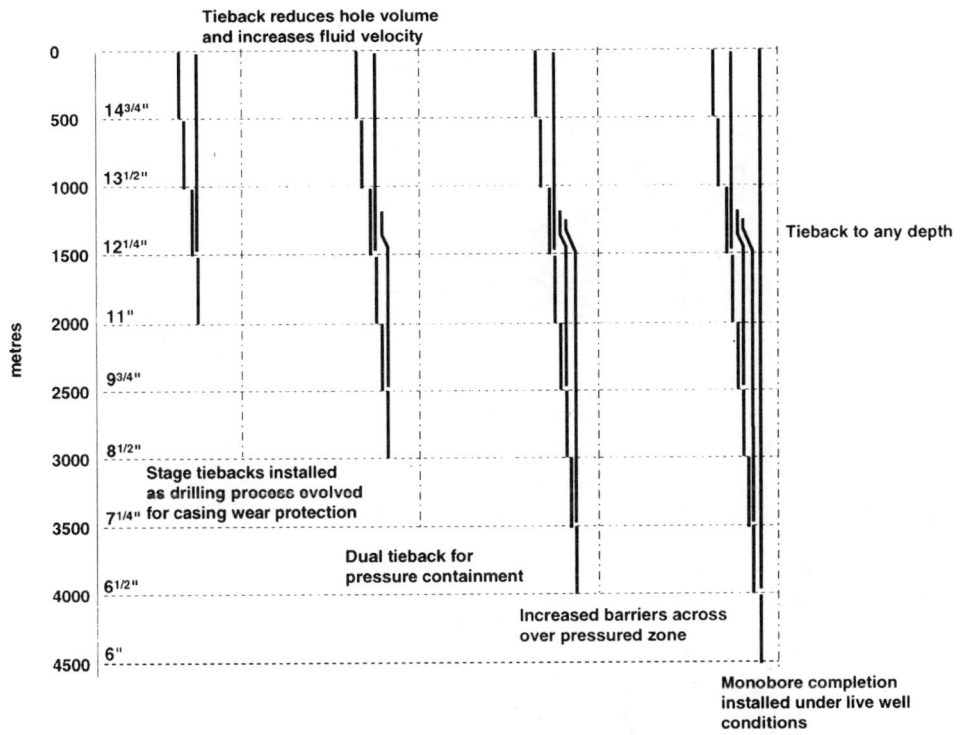

Figure 4. Liner and Completion tiebacks

MATERIAL SAVINGS/CONSTRUCTION ECONOMICS

Figure 2 proposed some alternative wellbore geometries. Using these dimensions a materials and costs spreadsheet was generated which provided the output summarised in the graphs below. These figures compare closely with other referenced work performed in this field[10]. Clearly, the slimmer the well geometry the slimmer the riser. Several studies have identified the upwardly spiralling cost of drilling conventional full size holes in deepwater. These studies have proposed reducing the riser size from conventional fullbore 21" to intermediate 16" and in the future 11" Clearly, the outcome of this project could have some very useful consequences to number of casing strings and or the conduit size across the reservoir.

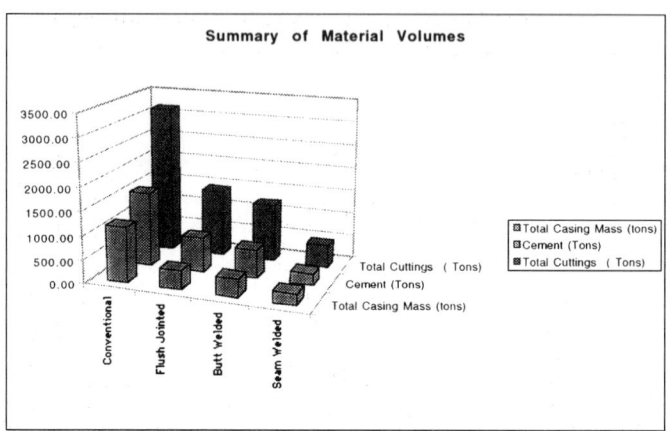

Figure 5. Using the wellbore geometries in figure 2 graphs show approximate material savings

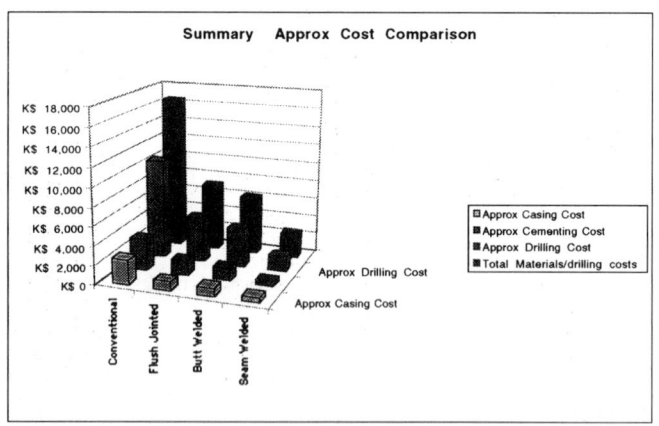

Figure 6. Using the wellbore geometries in figure 2 graphs show approximate cost savings

LINER INSTALLATION OPERATION

The major obstacle to overcome while installing the liner system described in this paper is that the very small annular clearances produce extremely high equivalent circulating densities (ECDs). This is turn can cause excessive pressure in the open hole section resulting in loss circulation or, in extreme cases total loss of wellbore fluids to a weaker formation. This has a considerable impact on well control issues, increased cost due to the loss of drilling fluids and potential damage caused to the reservoir itself.

These drawbacks are overcome by the equipment developed in this project. This comprises an innovative liner running tool and float collar/shoe which are connected by a flow tube (stinger tailpipe). This combination provides two circulation paths while conveying the new liner into the well, and a third circulation mode when it has reached its setting depth.

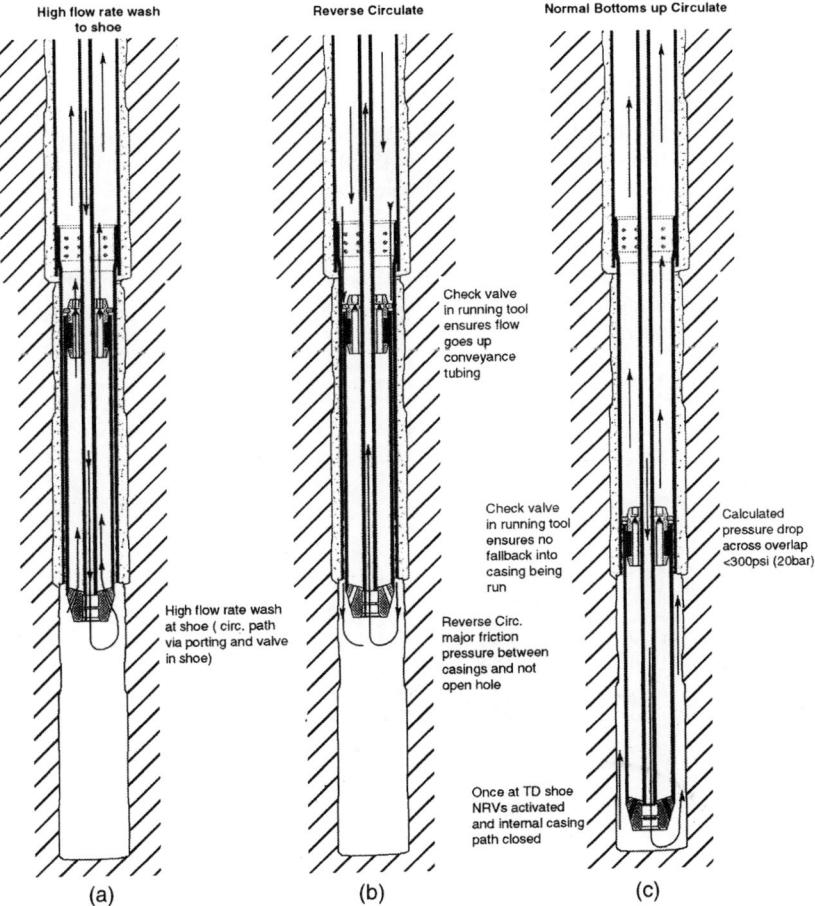

Figure 7. Circulation Options While running in the hole and when at TD

PERMANENT AND CONVEYANCES HARDWARE

Shoe design and operation

Figure 8 shows the general assembly of the liner shoe in its various modes of operation, and after it is drilled out and as it used to locate and secure the subsequent liner. Circulation modes shown in figure 7 a), b) and c) correspond with figures 8 a), b), and c).

Referring first to figure 8. a) the high circulation rate to wash and clean in front of the shoe is achieved by circulating the returns through unique porting back inside the liner being run. If it becomes necessary to circulate down the open hole, ports in the conveyance tool have check valves which force the fluid down the small annular gap between the previous casing and the new casing being run. Once the fluid reaches the shoe, it cannot pass into the liner being run because of the action of the check values in the running tool. The fluid can only flow up the inside of the tailpipe and conveyance tubing, eventually returning to the mud system a reverse manner. It will be appreciated that this mode of operation suits CT type applications more that jointed pipe as there are no breaks in the flow path to the mud system.

Finally, once the liner has reached the necessary setting depth, a ball is dropped which lands and seats on the sleeve retaining the shuttle valve in its circulation position and keeping the non-return valves open. Internal pressure is applied from surface and, at a predetermined pressure shear pins are activated which allow the shuttle valve to move to the closed position. The sleeve continues moving downwards where it is stored in a catcher sub. This process allows dual non return valves to become active for the cementing operation.

Eventually, once the shoe is drilled out, an internal profile remains which is used to both secure the subsequent casing and provide seal surfaces for the swaging operation which occurs during the liner hanger operation.

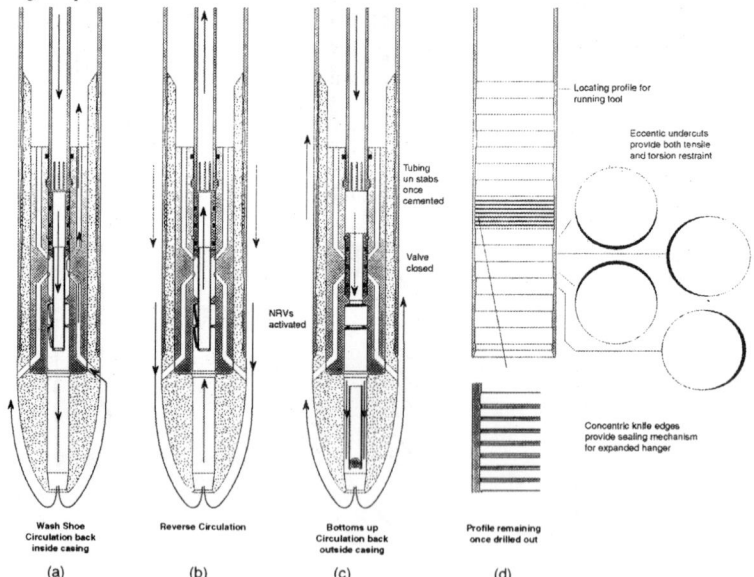

Figure 8. How circulation options are achieved at the shoe

High circulating while running in hole - figure 7a), 8a) and 8a)

A high rate circulation flow path exists to the shoe (via the conveyance tubing) through the tailpipe connecting the running tool to the float shoe. The return path passes through ports in the float shoe to the inside of the liner and then via ports in the running tool back into the annulus around the conveyance tubing. This circulation path provides excellent hole cleaning capability, while exerting minimum circulation pressure on the open hole. Although it is not shown the liner can also be rotated while being lowered into the well and, if used in combination with a reaming shoe[9], the risk of not getting the liner to its correct setting depth is greatly reduced. The flow ports discussed above as a similar flow area to existing NRVs and all functional surfaces not exposed to circulating fluids

Reverse circulation mode - figure 7b),7b)

While the liner is in the open hole, circulation may be required around the outside of the liner to provide additional lubricity. Because the annular gap between the previous liner and the new liner is very small, a high circulating pressure would be experienced. It would be unwise to circulate "bottoms up" as this would generate excessive circulating pressure across the open hole. A reverse circulation mode is possible with the developed tool assembly. Fluid is pumped into the tubing annulus. When it reaches the running tool the ports, which are used for the high circulating path incorporate check valves which force the fluid down the small annular path between the previous and new liner. Once the fluid reaches the open hole section the friction pressure drops significantly due to the hole being underreamed. The fluid then passes through the float shoe back to surface via the tailpipe and conveyance tubing. Again, the casing may be rotated during this operation.

Conventional bottoms up circulation - figure 7c), 8c)

Once at TD a different circulation mode is required. To convert the assembly to normal bottoms up circulation, a ball is dropped from surface. The ball passes through the running tool and lands in a sleeve in the float shoe. Internal pressure is then applied which converts the shoe to normal bottoms up circulation. The hole can then be conditioned at normal flow rates and the cementing process continued in a conventional manner.

Cementing Operation.

Because of the minimal clearances involved traditional mechanical casing centralisation cannot be employed. Advances in cementing slim casing annulus have progressed in recent years and these developments will be employed in this system. In addition, several novel non intrusive methods of achieving a 100% cement sheave will have completed lab demonstrations, and the most promising will be incorporated into the initial trails.

Conveyance tool and liner setting operation

This tool performs several functions as follows:

- supports the liner being conveyed into the well
- provides circulation modes during the conveyance and cementing operations
- provides surface indication of correct landing position.
- incorporates the securing and sealing mechanism.

Finally, it can be released both in its designed manner and in an emergency override mode to allow safe recovery to surface.

Figure 9 a), b), and c) show the liner conveyance and setting tool in its various modes of operation. Technology has been adapted from production systems to provide a positive location while running the liner to TD. Once at setting depth, the liner may be rotated during the cementing operation as required. The main features are the circulation port with check valve used for high rate circulation mode, swaging expanders for energising the metal to metal seal, and dimple type indents which deform the liner hanger into eccentric grooves cut into the previous shoe. This dimpling method is a successful and established development in coiled tubing connectors. It offers excellent results as a method to secure liners together where the space is so small that the moving parts of a conventional hanger could not be used successfully.

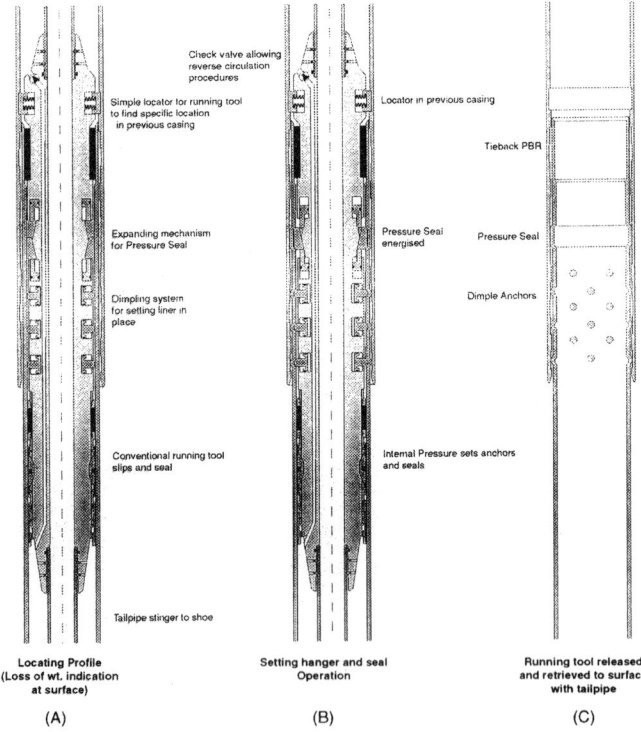

Figure 9. Running tool at TD, while circulating, when setting liner and when POH

SYNERGY WITH OTHER INDUSTRY DEVELOPMENTS

With the increased focus in activity in deepwater developments, ways of reducing exploration costs are being investigated and positive steps are being taken to reduce the riser diameter from the existing fullbore 21" systems to 16" and in the longer term 11" and even 8" CT risers are being considered. In its coiled casing form, this project offers an ideal link using readily accepted technology for immediate use in the deepwater exploration market. Figure 10 and 11 below shows a reeled riser vessel currently under development for CT well intervention applications using a 6" OD riser upgradable to 8" diameter.

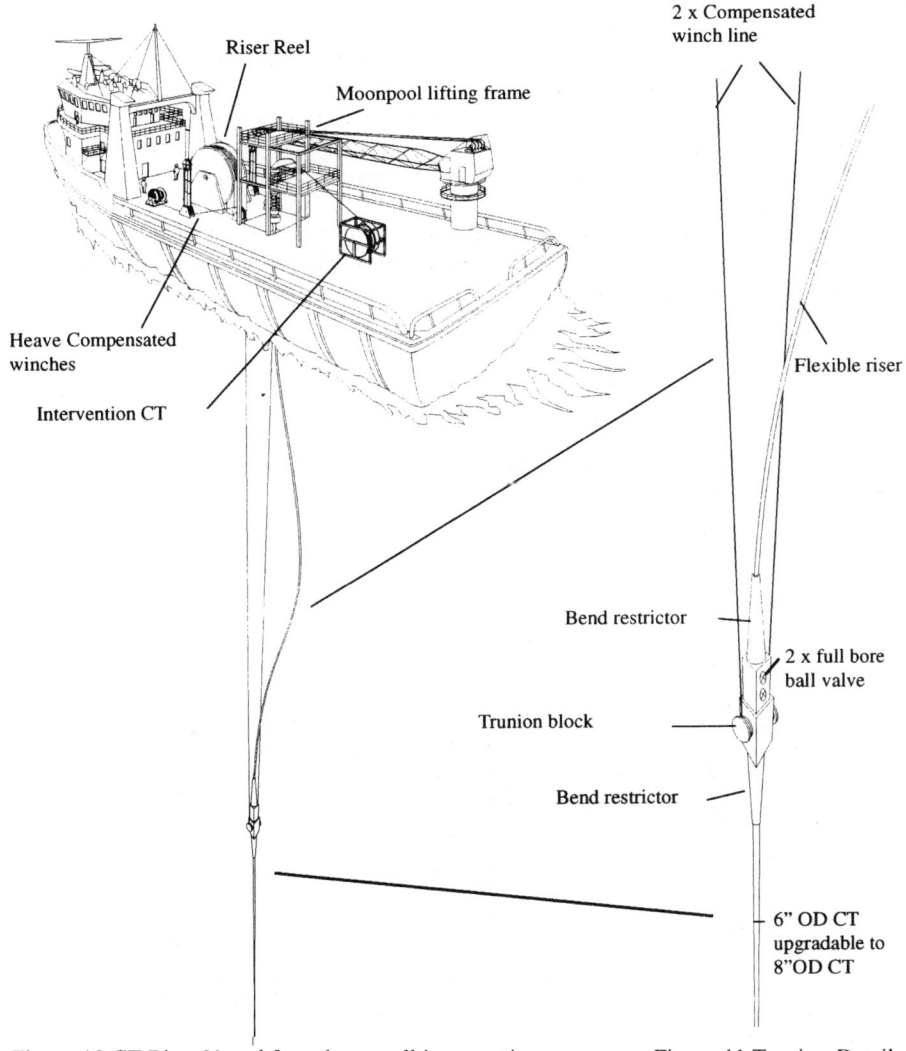

Figure 10 CT Riser Vessel for subsea well intervention   Figure 11 Trunion Details

SUMMARY

At the time of writing this paper the equipment required to demonstrate the effectiveness of this technique was in various stages of manufacture and test. The results from bench testing will be presented to the conference together with plans for a full scale demonstration.

The tools and techniques presented in this paper should provide the opportunity to slim the wellbore architecture from traditional dimensions while avoiding the high circulating pressures associated with small geometry wells.

With the increased focus on deepwater drilling and the slimming down of risers, the results from this project could offer considerable advantages for reducing well costs without sacrificing hole size across the reservoir.

The main advantages of the system are:-

- A conventionally sizes bore is maintained through the reservoir.
- Reduced logistics, handling and disposal costs
- Simple, fool proof and effective liner hanger system
- Significantly reduced well drilling and completion costs
- This innovative development now allows a broader range of options for well architects.

ACKNOWLEDGMENTS

The authors wish to thank the support and permission of the supporting companies to present this paper. These being ENI Agip SpA, Amerada Hess, Enterprise Oil plc, Saga Petroleum ASA and Shell International.

REFERENCES

1.R.N. Worrall et al. "An Evolutionary Approach to Slim-hole Drilling Evaluation and Completion" SPE24965 Europec 1992 Cannes France

2. E Carstens et al "Slimhole Drilling and Well Intervention from a Light Vessel" IADC/SPE 35043 IADC/SPE Drilling Conference 1996 New Orleans USA

3. Roger W Jenkins "New Concepts Lower Deep Water Drilling Costs" SPE30466 SPE Annual Conference Dallas 1995

4. E Kroell et al "Slimhole Completion and Production - What to do after we drilled the well" IADC/SPE 35129 IADC/SPE Drilling Conference 1996 New Orleans USA

5. B R Ross et al "Innovative Slimhole Completion" SPE24981 Europec 1992 Cannes France

6. C E Robison "Monobore Completions for Slimhole Wells" SPE 27601 European Production Operations Aberdeen 1994

7. Erwin Kroell et al "Slimhole Completion and Production" JPT September 1996 p846-847

8. R Cooper et. al. "Novel coiled tubing workover riser for cost effective well intervention" 2nd International Conference in Riser technologies. Aberdeen 25-26th June 1997

9. Reamer shoe Offshore Engineer April 1998 page 50

Bringing permanent reservoir monitoring technology and experience into new downhole applications

H HANSEN
Subsurface Technology AS (SubTech), Norway

INTRODUCTION (Fig 1)

For many years now, P&T (pressure and temperature) sensors have been installed in a majority of the wells in the North Sea. The system reliability as well as the personnel experience have substantially improved, since 1987 when better equipment was introduced - Designed for permanent downhole applications, not just short-duration downhole logging systems being adapted for permanent installations.

Subsurface Technology AS (SubTech) was formed in 1992 by personnel who designed, developed and successfully installed their first permanent gauge systems in 1987. Since the forming of the company, SubTech has followed the intentions to be an innovative company providing more complex well systems. These systems are today often referred to as "Intelligent" well completions.

In 1994, we installed 6 separate P&T sensors in a well, using one only cable to surface. All these sensors were addressable, and they all submitted data to surface within the same second. This was an important step forward, by proving that more than one sensor could be installed in a well using one common electrical mono conductor cable. We also proved that we could send commands to downhole electrical devices. Hence, being able to operate other addressable units (sensors, motors, valves, etc.) downhole was fully feasible.

This document will give an introduction where we are today regarding "Intelligent" well completion systems and where we will soon be with respect to further advances in this technology. We will have "Intelligent" well completion systems available ranging from simple systems to highly advanced systems where electrical equipment can be replaced by subsea rigless interventions, without having to pull the entire well completion.

A summary of our current activities with respect to multi-sensor system installed in a temporarily P&A (Plugged and Abandoned) monitoring well is also included; 132 sensors in one well, all transmitting data through the same cable. Some thoughts regarding the next generation systems for downhole monitoring in observation wells will also be shared.

CURRENT STATUS (Fig 2)

The permanent downhole gauge systems being provided today are improving in reliability. However, the most important aspect is that the personnel utilized for the installations in the field must be experienced, focused, dedicated and not at least; Proud of their work and what they produce!

There is no point in having a super-reliable system with super-electronics and super-sensors, if the personnel offshore are not super-focused on getting the system downhole in working condition and with the best foundation for a long and successful operation. Therefore our philosophy is that personnel doing the installations must be involved from the very start of the projects, to get the "This is the results of my work being installed now!" attitude. Hence, a suggestion to other vendors of such systems is to reward the personnel for innovative and creative thinking as well as for successful work.

Historical trends in the industry indicate that a significant portion of the failures in downhole permanent gauge systems has mainly been related to cable and termination failures. This is not our experience.
We have only seen one failure that can be proven to be the cable, and that has been caused by physical damage after the well system had been installed and set into production. We have not experienced any failures on connections, which is most likely the result of a substantial test program run at the start of the company which provided us with connections with double and independent sealing technology, as well as tension absorbers. Also, welding is now implemented on all couplings where applicable, as e.g. cable to gauge connection, cable to wet mateable electrical couplers, etc.

If the well casing program is evaluated ahead of the final design of the permanent gauge system, proper actions can be taken to prevent damage to the system while running the well completion. Such actions can be sufficient and well-designed cable protectors, protectors across tubing x-overs, protection of the cable across side pocket mandrels, etc.

What we observe though, is that electronic devices do fail and will continue to do so. No matter how much such units are tested and inspected, failures will occur sooner or later.

Complicated cooling methods can improve the survivability of electronics at elevated temperature, but it complicates the system and therefore also provides more components that will eventually fail downhole.

(Fiber optical sensor systems have now also been successfully installed in many wells, world wide. These systems has the advantage of having no active parts downhole. However, these systems can not provide any power to downhole tools as e.g. valves, solenoids, etc.)

Therefore our focus to this problem with the electronical systems is to provide more reliable electronics which is extensively design tested as well as properly (and for a sufficient time) tested at the conditions present in the well where the system is to be installed. Another improvement to system reliability is to provide a method where the electronics can be replaced without having to pull the well completion, when the electronics fail.

The SPGM system (Fig 3)

To obtain maximum lifetime on our downhole electronic systems, we have developed, tested and subsea installed systems where electrical and electronics can be replaced downhole, without having to pull the well completion to surface.

An advantage of such retrievability is also that upgraded components can be installed, whenever the customer wants to, when new sensors become available, when smarter downhole units has been developed, etc.

The retrievable method selected is based on a side pocket mandrel having a bottom-entry pocket for the electrical device (e.g. P&T sensor). Several advantages are provided using the bottom-entry configuration:

- No jarring required to install an electrical unit in the pocket (Straight upward pull installs the unit, while a hydraulic release running tool is utilized) - Minimum physical strain on sensors etc.
- Sensor pocket in upper section provides minimum risk of debris build-up on electrical coupler
- No interference with kick-over tools for conventional gas lift mandrels, etc.
- Electrical cable is fed straight into electrical coupler in top of mandrel; No bending of cable required
- It's easier to replace the component in the pocket by Well Tractors

THE WAY FORWARD (Fig 4)

We see that there is a need for more sensors and more "intelligence" downhole. Now the industry is finally starting to accept to install sensors in the wells, and now it is proven that such sensor systems can also be considered as fairly reliable equipment.

The SPGM system has now been further developed, and we now have a fully adjustable valve incorporated into the retrievable pressure- and temperature unit. This valve, having 2400 positions from fully closed to fully open, can be used for gas lift applications, as a circulating valve and as an inflow control valve for wells with lower production rates. The unit then contains a pressure sensor monitoring tubing pressure, a pressure sensor monitoring annulus pressure, a temperature sensor as well as the adjustable valve.

A large number of individual units can be installed downhole, all using the same electrical cable to surface. The units are addressable, which provides the operator to operate any valve or monitor any sensor connected to the system downhole.

A combination of retrievable sensors, retrievable valves, permanently installed sensors and valves can be installed within the same well.

The motor system incorporated into the retrievable valve can also be used for operating a ported sleeve which can be used for routing hydraulic oil to e.g. hydraulic operated inflow valves. It can also be used to extend or retract sensors or monitoring probes (e.g. for solids monitoring) into the well stream, from their standby position in the mandrel pocket.

We have filed a patent on a method where the sensor (or any device) in the SPGM can be replaced on a subsea well, without the use of a drilling rig. The operation can be performed from a small low-cost intervention vessel, and it is not utilizing subsea wireline, coiled tubing or well tractor run from the surface vessel. It is simply based on pumping the tool string into the well, and using the production pressure to transport and operate the tool system.

Abandoned well monitoring (Fig 5)
In June 1998, we installed a multisensor system in a subsea well offshore Norway at a water depth of 350 meters. This was a result of the work we have been doing for many years, where the methods of safely abandoning wells with sensors installed have been focused on.

The system consist of 132 sensors installed downhole, using one only electrical cable to the seafloor installed ROV replaceable datalogger. Hydroacoustic telemetry is used between the subsea datalogger and the surface vessel, for data uploads, reprogramming, status reports, etc.

The system are used for monitoring pressure and temperature downhole at various depths, as well as it will very accurately monitor the various fluid levels.

Further advances for the monitoring of abandoned wells are possible through the *patented* new methods for plugging and abandoning of wells we have launched. These methods provide increased reliability of the downhole barriers, the possibilities of abandoning the wells permanently and methods to secure the wells without a surface drilling rig system.

We have patented methods for platform- and subsea well monitoring methods, using tubulars (e.g. production tubing) or cables "cemented" into the well bore.

One of these methods will be utilized in conjunction with a major field cessation project offshore Norway, where reservoir monitoring as well as plug integrity monitoring is requested.

Also, a method where wireless two-way telemetry between the sea floor/surface and sensors permanently "cemented" into the well is being developed.

"Intelligent" well completion systems (Figs 6–8)
In 1993, SubTech started the development of a valve to be used for inflow control. The purpose of this valve is to be able to, from surface, open and shut off the production from individual zones downhole without well intervention. The valve can also be used for controlling the injection of fluids and gas into zones.

Having already installed a system downhole where six individually addressable sensors were installed at various depths, we knew that we would also be able to install the sensors required to monitor such a valve. Therefore, the next natural step for us was to further develop the experience obtained toward an "intelligent" well completion system combining valves and sensors.

Basically, the systems we are developing consist of many valves downhole combined with the existing permanent gauge technology and know-how.

We have now the following systems available/under development:

1. Hydraulic system
   Inflow control system, where up to 4 individual inflow valves can selectively be operated from surface, using one common hydraulic line only Sensors can be installed, based on the standard sensor systems already available

2. Electro-hydraulic system
   A system where numerous valves can be selectively operated from surface, using one electrical cable and one hydraulic line. Contains pressure- and temperature sensors, permanently and/or retrievable types.

3. Fully electric system
   A system where numerous valves can be selectively operated from surface, using one electrical cable only. Contains pressure- and temperature sensors, permanently and/or retrievable types. Can also operate the downhole safety valve(s) and production packer(s) electrically, via same cable.

Looking into the near future, we will see much more parameters being monitored downhole. There are many sensor manufacturers already providing or developing sensors such as flow meters (also multiphase types), sand probes, chemical sensors (for scale prediction, etc.), seismic sensors, etc.

To incorporate already at this stage a method where the present sensors can be replaced by these new sensors at later stage, without having to pull the well completion, is designed into the "intelligent" well completion systems we are developing.

Systems for downhole separation are also beginning to appear in the market. These systems also require highly advanced monitoring as well as surface operation of downhole valves for inflow and other fluid movement requirements.

Further ahead the "intelligent" well completion systems may become truly "intelligent"? It can be envisaged that the downhole sensors, the valves, etc., are coupled to a computer system onshore where interactive reservoir management is implemented. Completely computer driven, the wells can adjust flow rates perform testing of individual zones, perform functional testing of downhole safety valves, adjust gas- and chemical injection rates, choke down and shut off water producing zones, etc.

Initially the systems will provide a plan for action to the responsible reservoir- or production engineer prior to performing downhole operations. However, as the system "learns" to "understand" the field and its behaviour and the operator gives authority to the system, the downhole systems can perform operations that are analysed and reported to the operator after being performed.

More functions will be introduced downhole, which will simplify the requirements to the subsea wellhead and control systems. We will see wellhead systems much simpler than today.

Fig 1

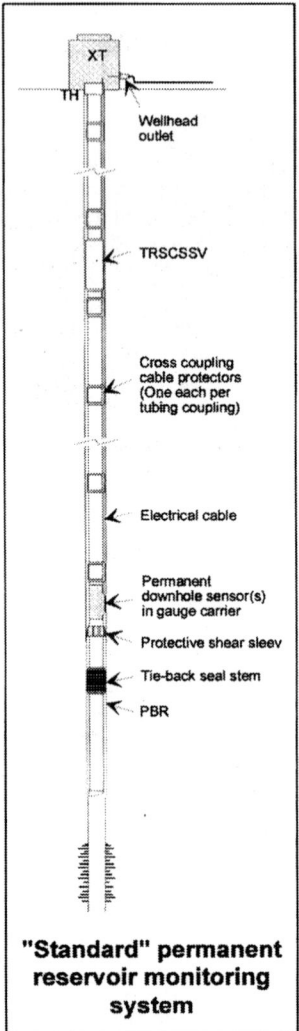

"Standard" permanent reservoir monitoring system

Fig 2

The people - Our most valuable asset

Fig 3

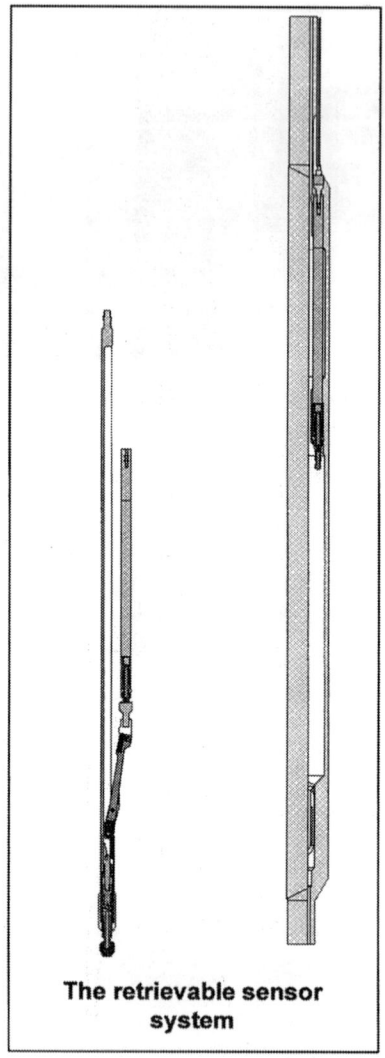

The retrievable sensor system

Fig 4

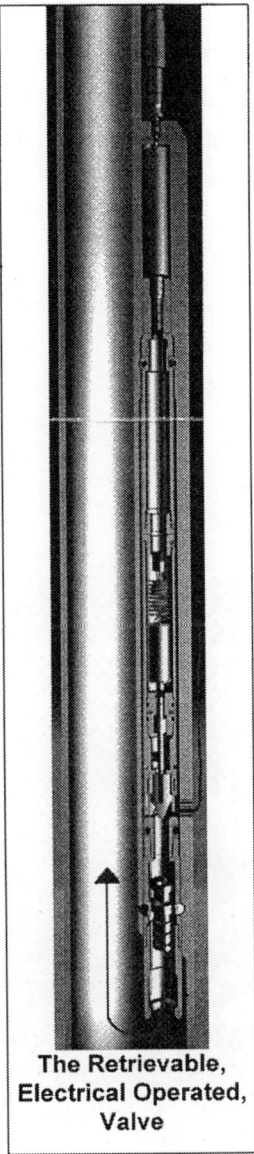

**The Retrievable,
Electrical Operated,
Valve**

Fig 5

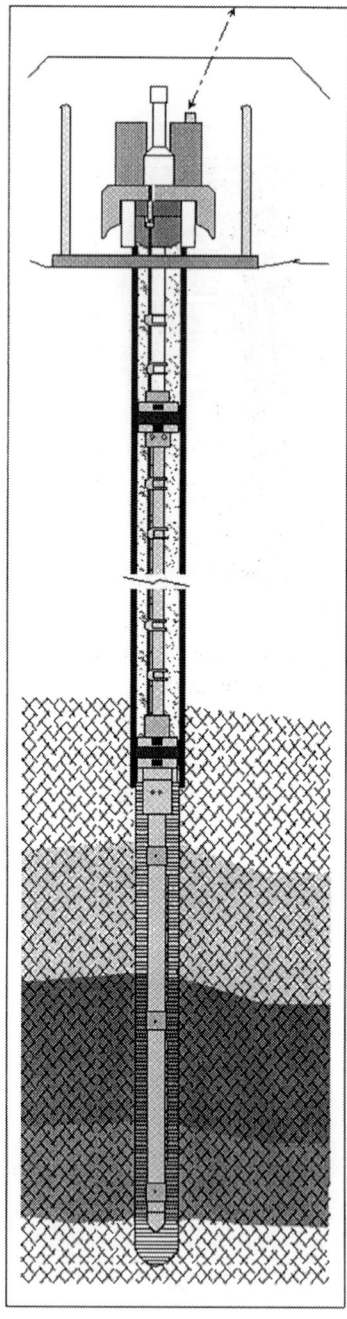

Fig 6

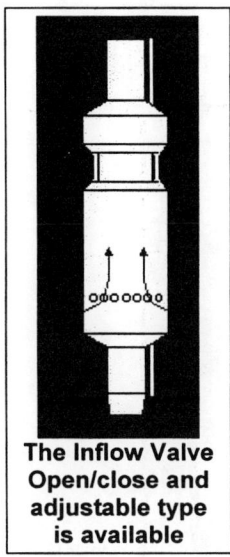

**The Inflow Valve
Open/close and
adjustable type
is available**

Fig 7

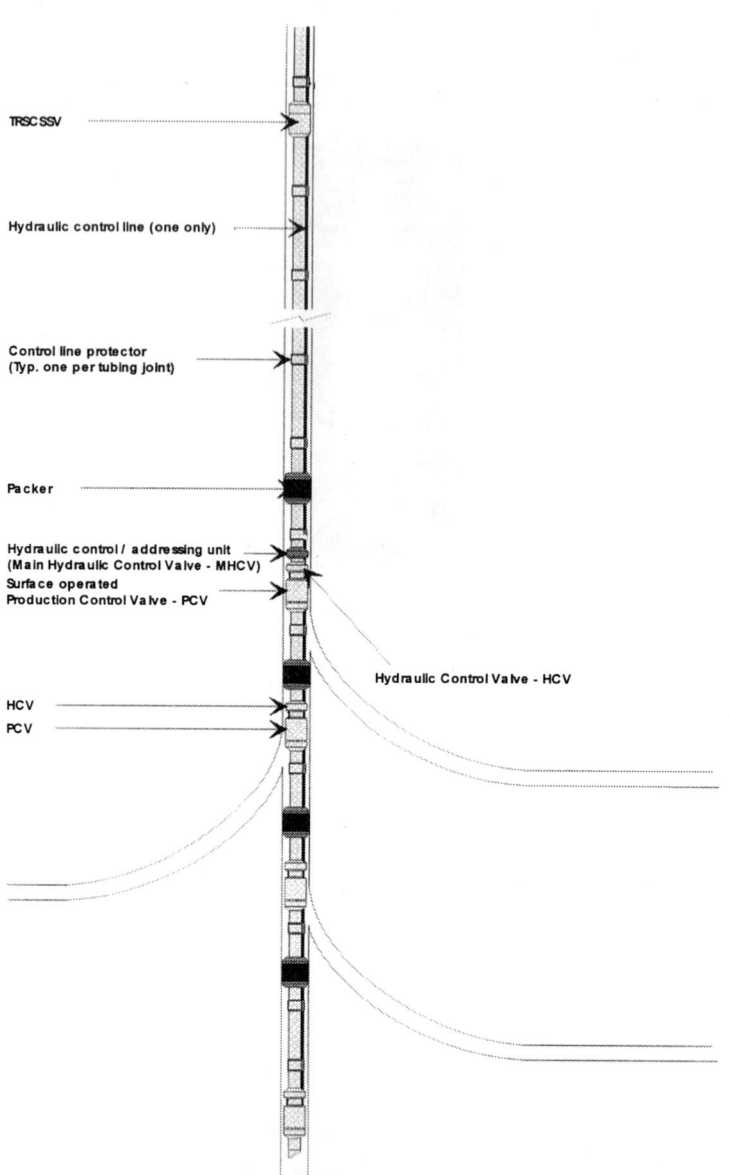

TRSCSSV

Hydraulic control line (one only)

Control line protector
(Typ. one per tubing joint)

Packer

Hydraulic control / addressing unit
(Main Hydraulic Control Valve - MHCV)
Surface operated
Production Control Valve - PCV

Hydraulic Control Valve - HCV

HCV
PCV

"Intelligent" well completion system - The hydraulic operated and addressable inflow control system
If reservoir- and/or inflow valve monitoring is required, our standard monitoring system can be install into same system

Fig 8

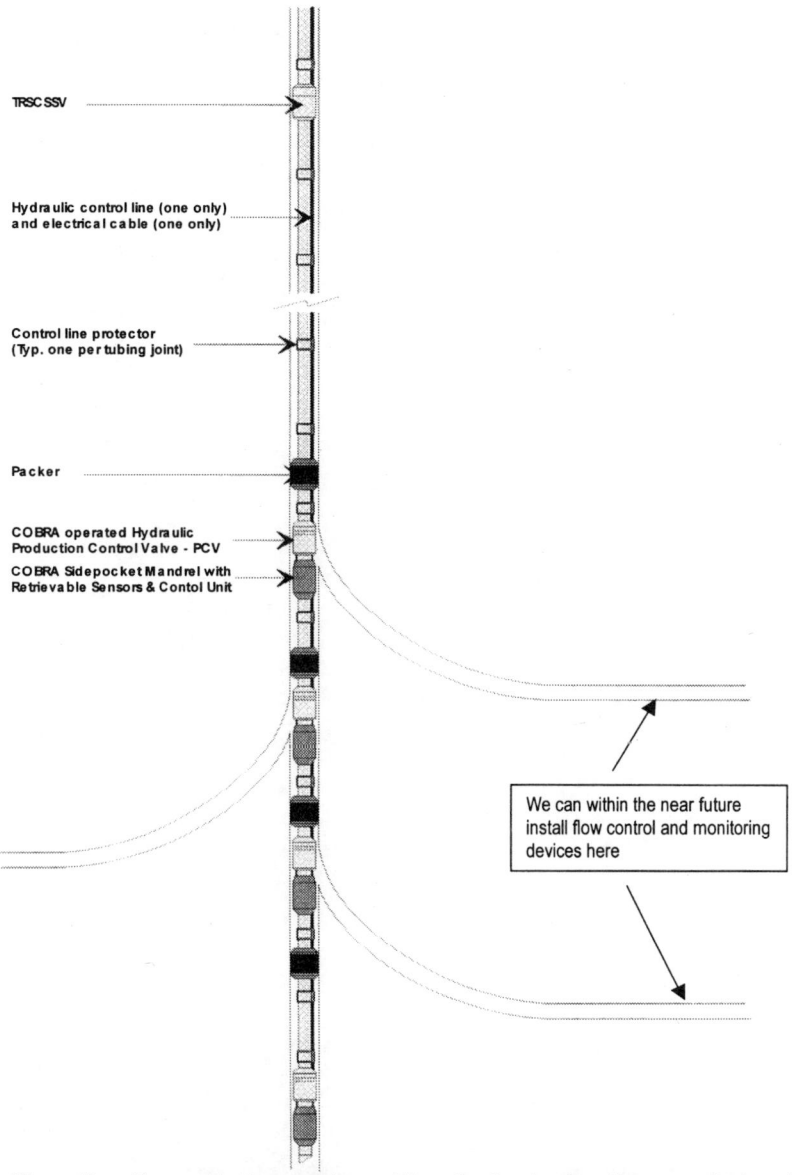

TRSC SSV

Hydraulic control line (one only)
and electrical cable (one only)

Control line protector
(Typ. one per tubing joint)

Packer

COBRA operated Hydraulic
Production Control Valve - PCV

COBRA Sidepocket Mandrel with
Retrievable Sensors & Contol Unit

We can within the near future
install flow control and monitoring
devices here

"Intelligent" well completion system - The electro-hydraulic operated and
addressable inflow control system
Includes reservoir- and/or inflow valve monitoring sensors
We can also provide fully electric inflow control system, where no hydraulics
is required from the wellhead

Listen to your reservoir – applications of microseismic monitoring

A J JUPE, J F COWLES, and **R H JONES**
CSM Associates Limited, UK

Synopsis

Microseismic events are caused by oil field production activities and can be located in space using seismic sensors deployed temporarily or permanently in monitoring, production or injection wells. This data can provide valuable information on geomechanical processes, particularly pressure distributions, occurring in the interwell region at distances beyond 1 km from the monitoring well. Microseismic monitoring fits with industry trends towards permanent monitoring and imaging remote from the borehole. The technology for temporary monitoring is well developed and field proven onshore and offshore in a variety of reservoir types. Permanent microseismic monitoring, particularly offshore, still faces technological challenges which may be overcome by extending existing technologies or by adopting emerging technologies.

Introduction

Microseismic events are very small earthquakes occurring on failure surfaces (e.g. fractures) which have radii of around 10 m. They are induced during fluid extraction and injection operations which cause shear-stress release on pre-existing fractures due to changes in pore pressure. *(Figure 1)*

Because the stresses acting on a rock mass are anisotropic, shear stresses build up on fracture surfaces. Under normal conditions these fracture surfaces are locked but when the in situ stresses are perturbed by reservoir activity, such as changing fluid pressure, the fractures shear producing small earthquakes. The seismic signals from these microearthquakes (also known as microseismic events) can be detected and located in space using geophones/accelerometers. This means that mapping the microseismic events is the same as mapping the location of pressure/stress changes within the reservoir. To date events have been located at distances exceeding 1 km from the wellbore in producing oil and gas reservoirs. The range of lithologies in which useful microseismic data has been gathered ranges from basement rocks[1] to chalks[2] and unconsolidated sandstones[2]

It has been demonstrated by a number of groups that microseismic events can provide substantial information about the distribution of fluid flow and pressure front movement within naturally fractured reservoirs, providing a practical and economic means of imaging flow and stress changes remote from the borehole.

The technology necessary to apply microseismic monitoring has been around for about 20 years in the geothermal, mining and underground storage industries but is only now beginning to be adopted in the oil and gas industry. The acceptance of the technique in the oil and gas industry is being driven by 3 main factors: (1) the desire to measure and monitor conditions and processes remote from the wellbore to understand hydromechanical behaviour in the interwell region (eg Deeplook[3]); (2) the refocusing of 3D reflection seismics from imaging geology to obtaining information on geomechanical and production processes remote from the well bore using time lapse surveys and (3) the move towards intelligent completions and the mastering of technical problems associated with permanent deployment of suitable seismic sensors.

Microseismic monitoring is a technique that can monitor certain hydromechanical processes and inferred geomechanical outcomes in real-time and at distances of >1 km radius from the monitoring well, making it a valuable and unique tool for achieving 3D spatial monitoring, providing information on:

Fault delineation/reactivation   *Compaction processes*
Fluid pressure front movement   *Flow path anisotropy*
Wellbore stability   *Hydraulic fracture strike, orientation and extent*
                                                        Cuttings re-injection monitoring

Fault Reactivation - imaging geological structures

Where faulting results in a compartmentalised reservoir, microseismic monitoring can help determine whether development/production related activity reactivates any of the faults and indicate both the fault geometry and adjacent pressure changes. There are examples of this application of microseismics from the oil and gas and geothermal industries where it has been used to determine the risk of premature watering out of production wells[2] and identify water injection losses to the far field[4] respectively *(Figure 2)*

Compaction

Compaction results in areas of localised failure which can be located continuously during the production life of the reservoir using microseismic events. This provides 4D information to assist in key development decisions, including placement of production/injector wells[4] and avoidance of wellbore stability problems.

Fluid pressure front movement

Pressure changes due to injection and production induce microseismic activity which means it is possible to carry out continuous real-time 3D monitoring of pressure front movement during injection or depletion *(Figure 3)*. This may allow un-depleted regions of the reservoir to be identified and decisions to be made about how best to exploit them.

Flow anisotropy

If the presence of fractures results in anisotropic flow, then microseismicity can be used to map out where the fluid is going. There are a number of examples of this in the literature[2,4]. This has significant impact on future well targeting[4], reservoir development, the planning of flooding, frac and drill cuttings re-injection programmes[5,6] *(Figure 4)*.

Integration of microseismic data with reservoir models

The link between microseismic event locations and fluid flow means that microseismic monitoring has significant potential to assist in the management of fractured reservoirs through the construction and then application of a reservoir simulator grid using microseismic data. It has already been demonstrated that microseismic events can be used in the construction and application of a reservoir simulator grid to production from a naturally fractured reservoir[4]. Work to date has shown that a 3D microseismic image of the reservoir can be used to obtain both the correct spatial distribution and correct hydrogeological properties of the simulator cells required to reproduce the observed anisotropic flow behaviour, the spatial distribution of the major flow paths and the tracer transport characteristics of the system. The value of the process is that a conditioned grid of cells can then input into sophisticated reservoir simulation packages to assist in predictions of long term reservoir behaviour and development.

This work undertaken so far confirms the importance of fracture connectivity in controlling the flow behaviour of naturally fractured systems and highlights the significance of microseismic imaging for 3D reservoir characterisation and monitoring *(Figure 5)*.

Current status of microseismic monitoring technology

The theory and processing techniques for dealing with microseismic data are well developed and proven. Current developments are focusing on improved location techniques[7] and integration of microseismic data with other datasets and simulators[4] commonly used in oil field development.

The hardware situation is less well defined. There are clamped tools available for short term survey deployments on wireline, slickline, on tubing and through tubing deployments. For land fields there are both temporary and permanent monitoring solutions available now with tools suitable for cementing bottom-hole or behind casing deployment.

For temporary surveys in offshore fields the lack of well availability and rig time remain and significant barriers to uptake of the technique, whilst for permanent monitoring solutions the costs of well intervention and cabling, data transmission and coupling problems mean there is still significant development work to do to create a turnkey permanent monitoring service. The low profile of microseismic monitoring in the oil and gas industry means that the business case is still to be made.

Temporary surveys

Temporary microseismic monitoring surveys are an essential step to establish the level and nature of the microseismicity and the type of information the data will provide on specific reservoir behaviour. A temporary survey will validate the design study network modelling and help determine the cost:benefit of moving forward to a permanent monitoring solution with an optimised spatially distributed sensor network. Such surveys can be achieved using a modified multi-level Vertical Seismic Profiling (VSP) tool deployed in an observation well within 1 km of the zone of interest[8]. Monitoring would generally be carried out for at least 5 days.

The monitoring sensors need to be deployed in a borehole as close to the zone of interest as possible. The microseismic events associated with oil field production and injection are typically of magnitude <0 and in anything other than an extremely shallow field would not be detectable at the surface due to attenuation. Microseismicity has a typical corner frequency of a few 100 Hz and may contain high frequency information that is subject to heavy attenuation losses, therefore for every event of magnitude >1, located at say 1 km, there may be 10 of magnitude zero and 100 of magnitude -1. A surface network would typically only detect events of a magnitude associated with tectonic activity (>1 M_L). The large volume of low magnitude microseismic activity is key to obtaining high definition images of structures, such as reactivated faults, and production related hydromechanical and geomechanical processes in the interwell region

The key steps in undertaking temporary microseismic monitoring surveys are, typically:

- Initial feasibility, modelling and design study
- Short term field survey using multi-level VSP tool (>5 days)
- Data processing and interpretation

Temporary survey examples

During April 1997 in the Phillips Petroleum Ekofisk field a temporary microseismic monitoring survey was carried out for 18 days using the CGG triaxial SST500 VSP tool with 6 levels deployed near the crest of the field. The event rate was ~100 per day locating approximately 2100 events ranging from magnitude -3.5 to 0 up to 2.2 kilometres from the monitoring well. The clusters of microseismicity located under the gas cloud indicated possible reactivated fault structures which could not be imaged using reflection seismics due to the presence of the gas cloud. The results of this survey are detailed in *Maxwell et al, 1998*[8].

The same tool was deployed for 60 days in the Valhall field for Amoco, Norway. The tool was lowered to the planned depth, roughly 200 m above the reservoir formation, and locked in position. The whole array then remained in position for 60 days operating continuously and with all data from the 18 seismic channels (six levels all 3-component) transmitted to surface in real time. At surface the microseismic data acquisition system isolated and stored the records where recognisable seismic signals were evident. The stored data were downloaded and sent to shore for processing and analysis. Initial results indicate alignments of the microseismic events which may provide information on the structures within the reservoir.

Further processing of the data will take place in Autumn 1998.

Permanent monitoring

It is important to note that permanent microseismic monitoring does not require a massive array of sensors in each well. The spatial distribution of the network is more important than the number of sensors. A reservoir can be monitored perfectly well with as little one single triaxial geophone in each of 3 wells. However, the fewer the wells instrumented and the poorer the spatial distribution of the wells relative to the optimum sensor locations, the more sensors are required per well.

The drive to image the interwell region and develop smart completions capability is focusing industry attention on permanent monitoring, requiring deployment of sensor packages as part of the well completion. For microseismic monitoring there are already some tools with the potential for behind-casing and on-tubing deployments, but there is still some development required. However, integrating seismic tools with smart completions faces significant technical problems.

1) Currently the industry is at the stage of pilot microseismic trials using existing technology. Typically this involves geophone tools cemented behind casing requiring special wireline type cables or downhole digitising and multiplexing of data to surface via monoline cable. The latter option is more complex. Digitising downhole is well established for VSP tools but requires some development to handle long term deployment and the larger microseismic data rates. The principle issues to be resolved, after the reliability of sensor package and cabling has been addressed, are decoupling the sensor package from well noise, effective coupling to the formation and cabling back through existing wellhead designs.

2) It is feasible to integrate seismic sensor packages within smart or intelligent completion systems but this requires going a stage further and implementing limited downhole processing due to bandwidth limitations. Rather than returning the full seismic waveform in analogue or digital form to the surface the data will be processed to capture the P & S wave arrivals and certain attributes necessary for further surface processing. Implementing limited downhole processing would probably require high levels of electronics integration to achieve the extended lifetimes, which in turn means long development lead times and extensive investment.

3) The long term future of permanently deployed seismic sensors may lie with fibre optic "accelerometer" sensor packages using Bragg gratings technology[9]. The thermal and bandwidth performance limitations of conventional systems are removed and reliability is increased by removal of complexity, particularly downhole electronics. As with conventional technologies coupling the sensor packages to the formation remains an issue, as does cable protection and connector integrity. The number of sensors that can be deployed per well is limited by choice of application and surface acquisition constraints rather than bandwidth. This opens up the potential for applications such as repeat 3D VSP surveys and permanent crosswell seismic surveys which require a much higher number of sensors in the well than microseismic networks.

Conclusion

Microseismic monitoring is an important technique which provides valuable information on hydromechanical processes remote from the well-bore. This information will impact production and field development decisions as it is more widely adopted and operators become more familiar with how to integrate it with current best practice. The full potential of microseismic monitoring cannot be realised until seismic sensors can be routinely deployed permanently in boreholes as part of the completion process. This requires a better understanding of the business case and investment in borehole tool development.

References

[1] Wallroth T, Jupe A and Jones R, 1996. Characterisation of fractured reservoir using microearthquakes induced by hydraulic injections. Marine and Petroleum Geology, Vol. 13, No. 4, pp447-455.

[2] Phillips W, Rutledge J, Fairbanks T, Gardner T, Miller M and Schuessler B, 1996. Reservoir mapping using microearthquakes: Austin Chalk, Giddings field, TX and 76 Field, Clinton Co, KY. SPE 36651.

[3] Stoessel E T, 1998. Access and integration of emerging technologies: keys to successful imaging of reservoir fluids in depth and time. Paper OTC 87161, Offshore Technology Conference, Houston, Texas, 4-7 May 1998.

[4] Jupe A, Jones R, Dyer B and Wilson S, 1998. Monitoring and management of fractured reservoirs using induced microearthquake activity. SPE 47315.

[5] Brady J, Withers R, Fairbanks T and Dressen D, 1994. Microseismic monitoring of hydraulic fractures at Prudhoe Bay. SPE 28553.

[6] Keck R and Withers R, 1994. A field demonstration of hydraulic fracturing for solids waste injection with real-time passive seismic monitoring. SPE 28495.

[7] Jones R and Stewart R, 1997. A method for determining significant structures in a cloud of earthquakes. Journal of Geophysical Research, Vol. 102, No. B4, pp8245-8254.

[8] Maxwell S, Young R, Bossu R, Jupe A and Dangerfield J, 1998. Microseismic logging of the Ekofisk reservoir. SPE 47276.

[9] Kersey A 1996. A review of recent developments in Fiber Optic Sensor technology. Optical Fibre Technology 2, pp291.

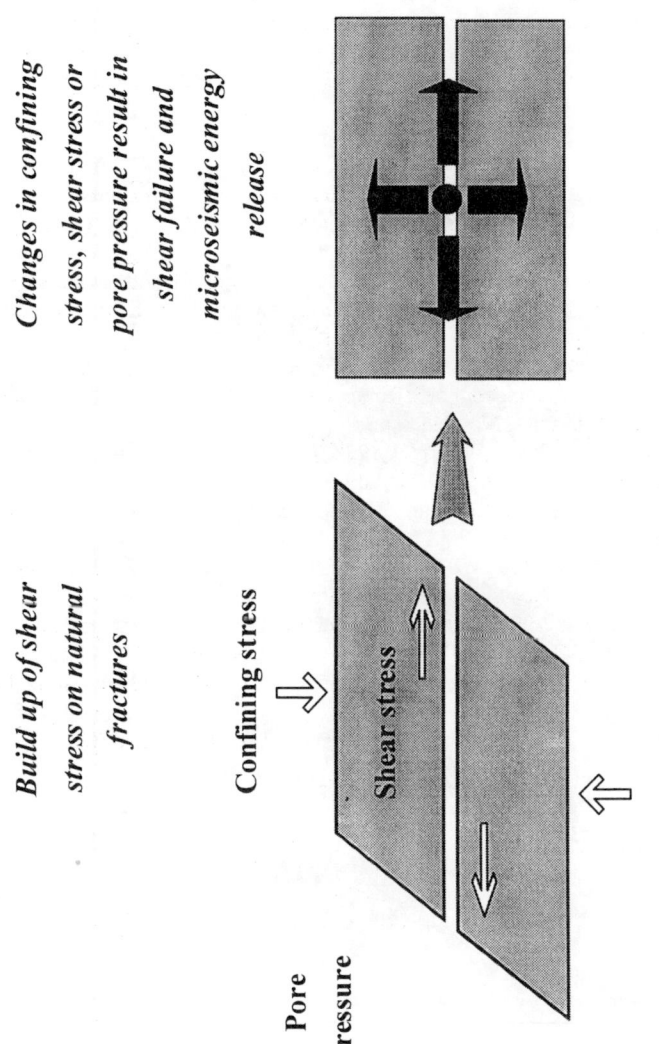

Build up of shear stress on natural fractures

Changes in confining stress, shear stress or pore pressure result in shear failure and microseismic energy release

Confining stress

Shear stress

Pore pressure

Figure 1: Conceptual schematic of microseismic source mechanism

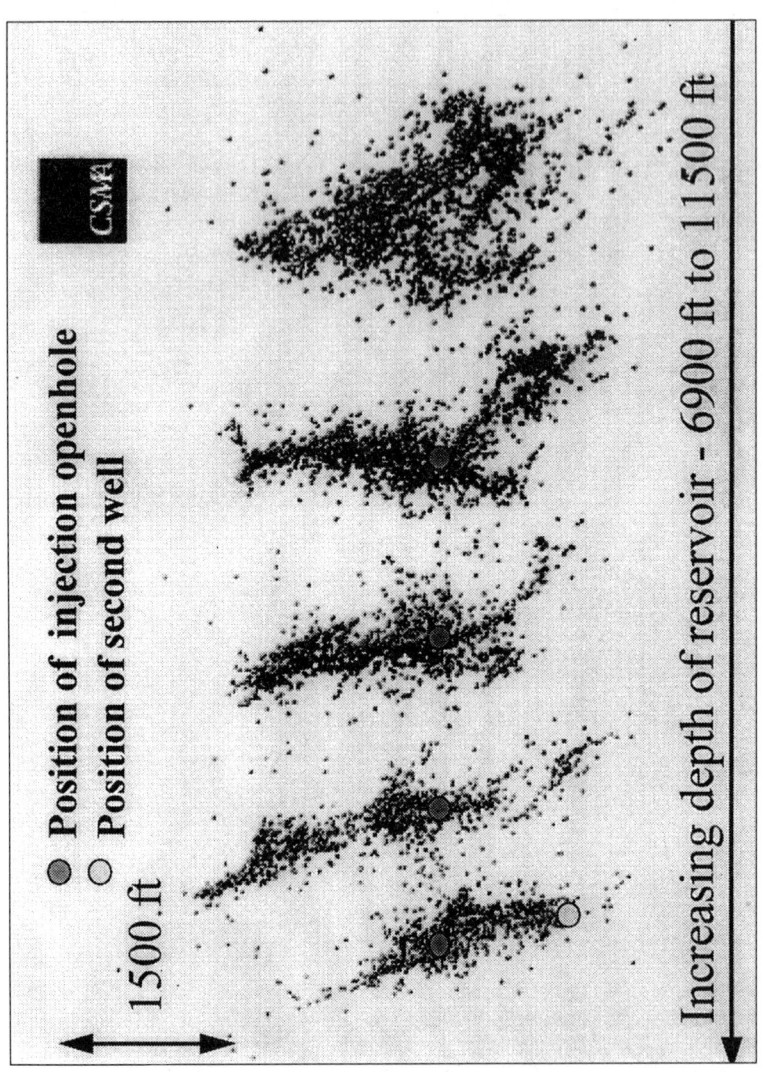

Figure 2: Plan view of 300 m depth slices taken through a cloud of 16,000 microseismic events detected during a water stimulation. The distribution shows strong flow anisotropy and evidence of defined structures.

Figure 3: Time-lapse image of microseismic events generated by pressure front movement moving out from the injection point during an hydraulic stimulation

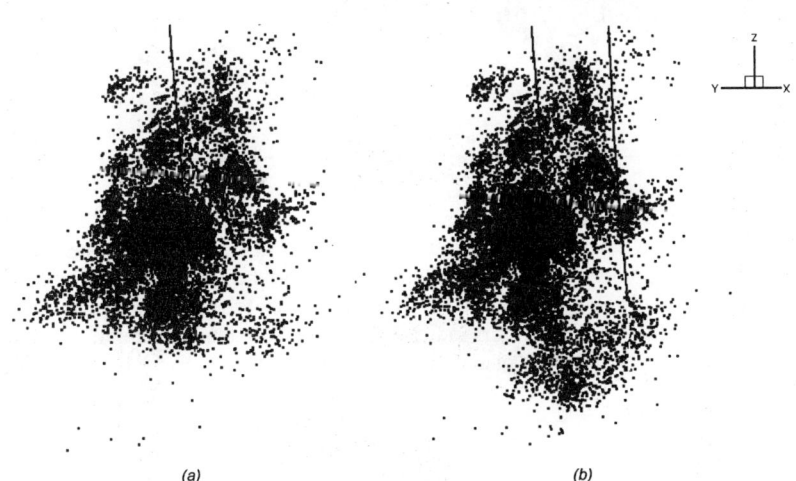

(a)   (b)

Figure 4: An example of microseismic data being used to target an injection well in a geothermal reservoir. (a) shows microseismic events generated during an hydraulic injection and (b) shows the injection targeted into the deepest part of the seismic cloud to ensure the highest temperature and good hydraulic coupling with the production well.

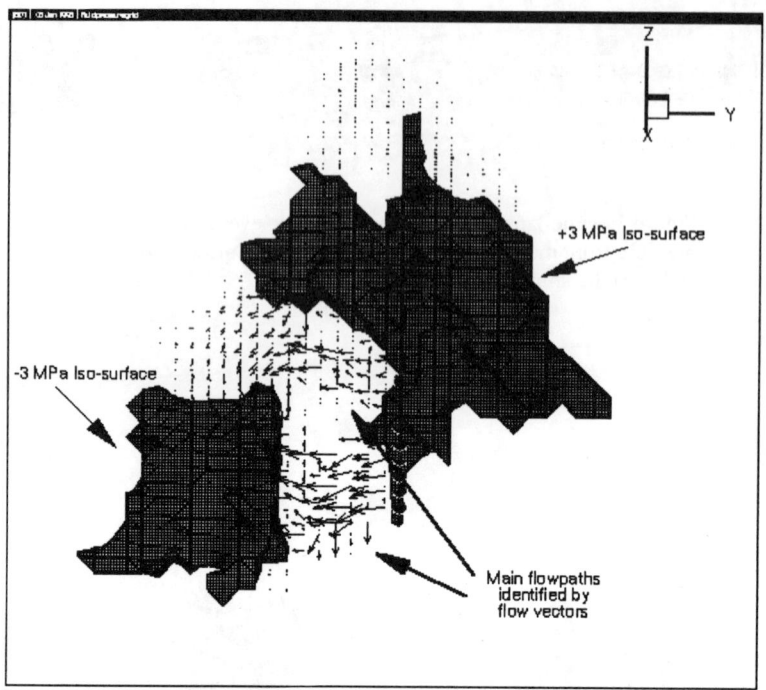

Figure 5: Steady-state fluid pressure isosurfaces and flow vectors for circulation between injector and producer wells in a geothermal system. The well nodes are shown as black spheres and the length of the flow vector arrows represent the relative magnitude of the flow velocity. Note that the flow vectors indicate discrete flow paths that are out of the plane between the two wells.

A configurable subsea water boosting and injection system

A RADICIONI and **G D'ALOISIO**
Sasp Offshore Engineering, Italy
A BAKER
Baker Jardine & Associates, UK

1   ABSTRACT

The economic exploitation of marginal fields, especially those in deepwater, requires the introduction of innovative and cost effective technologies .

"Subsea processing" is one of the challenging new technologies that can provide significant cost savings which help facilitate the economic exploitation of deepwater oil and gas reserves. Subsea processing improves field economics by boosting production from the reservoir and/or reducing topside processing facilities.

In order to achieve maximum flexibility while complying with the Operator's requirements, a "building block" approach has been adopted based upon a number of technologies which have been developed based upon various aspects of subsea multiphase boosting and subsea separation.

One of the main building blocks is subsea water boosting and re-injection into a reservoir. This is the subject of this paper.

The proposed system can be used as a part of a subsea processing system to enable the separated water to be injected in the reservoir to avoid costly transportation and surface processing infrastructure (the base case for this paper). Alternatively, the system can be configured for raw water injection in order to maintain the required reservoir pressure and maximize production.

The concept is characterised by the maximum use of field proven equipment which can be easily reconfigured during the field life to follow changing production profiles. The system engineering design has now been completed to a high level of definition such that detailed design and fabrication could be initiated without the requirement for further conceptual or front-end engineering.

The necessary water injection pressure is provided by Electrical Submersible Pumps (ESP's) contained within a dedicated caisson which provides the required Net Pressure Suction Head (NPSH) for the pumps and is deployed vertically into a conductor set into the seabed. The proposed caisson concept offers a great deal of operational flexibility as the pump sets can be quickly and economically exchanged to match changing flow and head requirements during the field life. The proposed pump motor sets are "catalogue" models which have been extensively field proven in similar water injection duties and downhole pumping.

The paper describes several possible subsea layouts and configurations for various application cases, namely: subsea water separation and water re-injection into the reservoir, to avoid transportation, or reservoir pressure maintenance utilizing subsea raw water treatment and re-injection. Installation and maintenance procedures for the pump caisson assembly are also described, as well as the main characteristics of the proposed system.

2   INTRODUCTION

New field developments are moving to ever deepwer waters and/or further from existing infrastructure. Under these conditions the conventional production concept based upon the utilization of large field platforms to accommodate all the required equipment and facilities for separation, pressure boosting and well testing tends to become less economic.

Subsea processing options can significantly improve field economics by moving significant amouts of equipment from the remote production platform subsea and closer to the reservoir. Such configurations are less affected by environmental parameters and facilitate the full exploitation of the productivity of the reservoir. In this regard "subsea processing" is one of the challenging new technologies that can provide significant cost saving and improve field economics by increasing production from the reservoir and/or reducing topside processing facilities.

In order to achieve maximum flexibility while complying with field development requirements, a "building block approach has been adopted based on a number of technologies which have been developed based upon various configurations of subsea separation, boosting and multiphase transport. Sasp/Eni Group have been active since the mid-eighties developing first, multiphase boosting, well testing and metering technologies and more recently subsea separation technology and can now offer a wide range of mature subsea processing options. The generic term of the proposed "configurable" concept is CoSSP (Configurable Subsea Separation & Pumping System).

The concept design philosophy and a brief description of the building block approach is outlined together with the description of the proposed technologies and equipment. The system architecture options for standard applications are also outlined. The water pressure boosting and re-injection system is described in detail by way of an application case study based upon a subsea water separation and re-injection system proposed for a field located in the North Sea. The paper closes with an outline of some possible applications of the selected "building blocks" relating to the subsea separation and subsea raw water injection.

3   CONCEPT DESIGN PHILOSOPHY

The CoSSP concept is based upon existing, proven and readily obtainable components and technologies. The following design philosophies have been adopted in order to meet the design objectives:

- All components must be field proven in similar applications.
- All key components must be available "off the shelf", preferably from more than one supplier.

- Design must minimize the number of "active" components. I.e. minimize the number of components which may require maintenance.
- All "active" components to be maintainable from a small to medium sized Diving Supply Vessel (DSV).
- Minimize life-cycle costs by reducing the need and frequency of intervention.

4 BUILDING BLOCK APPROACH

The building block approach adopted in the CoSSP concept enable wellhead separation and pumping offering: extended reach by pressure boosting of high density phase only; enhancement of well productivity at increased water depths; stable flow regime in production flowlines. In addition CoSSP enable subsea water removal offering: reduction in total flowrate to be transported; reduction in power requirements for pressure boost; no bottlenecks on receiving treatment facilities; and mitigation of hydrate formation risk.

The separation technologies considered in CoSSP are conceived to enable both two phase gas/liquid separation and three phase separation: Gas/liquid separation is implemented by using a novel compact vertical separator (VASPS) with integral pump for liquids pressure boosting. Conventional three phase separation is adopted to remove the bulk water form the well stream and to make it suitable for either re-injection or disposal to sea. A number of export options are also possible, including combined oil and gas export with or without pressure boosting and separate oil and gas lines with oil pressure boosting. Three phase well testing and metering is also integrated into the system.

Boosting and injection, depending on wellstream process conditions, can be performed by electrical submersible pumps or by subsea multiphase twin screw pumps, which have been developed together with Nuovo Pignone. In both pumping systems the flowrates are regulated through a variable speed driver (VSD).

Considering deepwater applications, diverless installation and intervention have been considered for the system design and module retrievability it is foreseen for those active equipment that require maintenance during field life.

The whole CoSSP concept is fully engineered and the main building blocks are constituted by technologies that have been qualified through prototype testing or by equipment that are "catalogue" models and can be supplies "off the shelf" by multiple vendors.

In the following sections the block of the CoSSP related to water boosting and injection is described in detail considering the two major applications related to subsea separated water re-injection in the reservoir, to avoid its transportation, or reservoir pressure maintenance, considering subsea raw water treatment and re-injection.

5 SYSTEM ARCHITECTURE OPTIONS

For the above mentioned applications different options have been considered and are available within the CoSSP concept depending on the design basis of the considered field and by the Operator's requirements. These options ranges from a subsea integrated configuration, where all subsystems/components are accommodated in a single template, to a more distributed subsea layout, where the equipment are arranged in two or more structures.

Figures 1 and 2 show block diagrams of possible subsea configurations; a comparison of the two considered applications is also shown.

Fig. 1 considers two architecture options related to integrated configurations and the main subsystems related to these subsea processing plants are shown in the figure; the water re-injection facilities are well identified and evidenced. The two architectures, for each applications, differ mainly from a control point of view and are in principle similar for both the applications.

The first case (Option 1) considers an integrated subsea control system able to monitor and to control in real time both the subsea processing plant and the water injection tree. The second case (Option 2), on the contrary, foresees two separate control pods: one standard for the water injection tree and another dedicated to the monitoring and control functions of the specific process (i.e. subsea separation or sea water treatment). In Option 2, the dedicated control pod is also related to the control of the water boosting system. A topside located supervisor control system integrate the information received by the two subsea pods and provides the overall process control both in operating or in start up/shut down conditions.

This last option is more flexible and it has the advantage of more standardised building blocks: the water injection tree and its related control system are completely standard and can be obtained by the suppliers without any specific requirements in addition to those imposed by the characteristics of any water injection well (flow rates, pressure requirements, water quality characteristics, etc...).

An evolution of the above described Option 2 is shown in Fig. 2 where a distributed architecture is outlined. The configuration shown in Fig. 2 is essentially constituted, for both the considered applications, by a subsea template on which all the processing equipment are located and the water injection well is physically separated from the template but it is connected from process point of view. The tree control pod is placed on the tree structure.

The same concept could be applied also for multiple water injection wells; in this last case the distributed configuration looks like as typical clustered wells: the water boosting system became more complex for the piping and valves arrangement and for the relevant control.

For the architecture of Fig. 2 the same considerations rised for Option 2 of Fig. 1 are still valid.

The control and power lines can be accommodated in two separate umbilicals or in an integrated power and control umbilical. In case the separated power and control umbilicals are selected, the tie-in porches are located closed to the relevant subsystem (water injection and subsea control system) and limited distribution facilities are necessary on the subsea structure. If integrated power and control umbilical configuration is selected the tie-in porch position is optimised considering the distribution in different locations, both of the power and of the control lines.

Other alternatives could be selected starting from those shown in Fig. 1 and 2.

For the subsea separation application, in addition to the re-injection of the separated water into the reservoir, two other alternatives have been developed namely: discharge into the seawater or transportation through a flowline up to topside facilities or to shore. These three possibilities mainly differ for the boosting requirements (i.e. pressure required) and for the water quality requirements; the most stringent water quality specifications are relevant to the discharge into the sea (pollution problems) and to the re-injection into the reservoir (integrity of the reservoir, i.e. porosity).

As shown in Fig. 1 and 2, the main building blocks to perform the required subsea processing are the following:

A) *Water separation and re-injection*
- water separation

- water quality monitoring
- water injection facilities
- control system.

B) *Sea water treatment and re-injection*
  - sea water treatment
  - water quality monitoring
  - water injection facilities
  - control system.

A) *Water separation and re-injection*

The water separation block is mainly composed of two or three phase separator and relevant internals and by the separator level control. This sensor (generally duplicated by two different types of instruments) is essential to assure the stable and good operating conditions and is part of the closed loop control of the whole system.

The water quality monitoring system mainly include the sensors for oil in water and sand content measure. Depending on the specific application and on the Operator's requirements such monitoring could be done continuously or by periodic sampling, taken from the subsea process unit by means of an ROV that is able to remove, to recover on surface and to replace the sampling system.

The water injection facilities, for both the considered application, better characterised in the following sections, mainly consist of: pump & motor sets, choke valve, water injection tree. The choke valve is required to regulate the water flowrate depending on the production parameters (i.e. production profiles of the production well).

B) *Sea water treatment and re-injection*

The seawater treatment is necessary in consideration of the stringent requirements to maintain the integrity of the reservoir and to reduce the risk of subsea processing equipment damages (avoid: components corrosion, encrustation, solid particles and fishes intrusion inside the water injection facilities, etc....).

The water quality monitoring system is required to check the water conditions after the treatment and before the re-injection process.

For the water injection facilities the same considerations outlined above are still valid.

Issues relevant to the overall control system architectures have been also discussed hereabove for both the considered applications.

6   CASE STUDY - SUBSEA WATER SEPARATION AND RE-INJECTION

The case study application described here below is a stand alone CoSSP template located adjacent to the well production template approximately 5 km from a production platform as shown in Fig. 3.

The function of the CoSSP template is to separate produced water from well production streams and re-inject the water into the reservoir. The hydrocarbon phases (oil and gas) flow back to the production platform under wellhead pressure. The production platform will provide control facilities, power and utility services for the CoSSP template. The CoSSP template will contain as a minimum: a separator vessel unit, a water injection pump, a well slot for a water injection well, a subsea control module.

6.1 System Design Premises

The main features to the CoSSP template system design are:

All equipment accommodated in a single compact subsea structure.

Design life:	25 years (for permanent equipment)
Water depth:	340 m
Max. electric power available:	2 MW at 11kV (on the production platform)
Installation/retrieval:	by guidelines or guidelineless
Subsea intervention:	by work class ROV

6.2 Process Design Premises

The main CoSSP template process design data are as follows:

- Number of producing wells : 5
- Max expected oil production : 3900 Sm3/d
- Max expected gas production : 810, 000 Sm3/d
- Max expected water production : 9200 Sm3/d
- Typical wellhead flowing pressure (WHFP) : 40 - 105 bara
- Typical wellhead flowing temperature (WHFT) : 50 - 65 °C

Separator Data

- Max oil content in the re-injected water stream (v/v) : 1000 ppm
- Water in oil content (reference value) : 10% v/v
- Typical pressure drop across the CoSSP template : 1- 2 bara
- Water retention time : 6 mins
- Oil retention time : 3 mins

Pumping System Data

- Target injection capacity : 6000 Sm3/d
- Max topside available power : 2 MW
- Water injection well tubing head flowline pressure (THFP) : 190 bara

6.3 Design Philosophy

The basic philosophy of the proposed process scheme is to maximise simplicity, reliability and the use of field proven technology and equipment, considering the design life of 25 years.

The main components of the CoSSP template are a single horizontal separator and a single retrievable pumping system. The function of the separator is to separate the produced water to injection quality (1,000 ppm oil in water or better). The function of the pumping system is to boost the pressure of the separated water to allow injection back into the reservoir.

Process Philosophy

The primary functions of the CoSSP template are:

- to handle the total well stream of the production flowline from the subsea production templates
- to provide an efficient water separation (better than 1,000 ppm)
- to re-inject in a well the maximum water flowrate allowed by the available power on the production platform
- to re-route the oil and gas stream to the production flowline.

The CoSSP template separator will separate the water from the wellstream. Produced water with a quality of less than 1,000 ppm oil in water will be directed to a pumping system where the pressure will be boosted to inject the water back into the reservoir via a water injection well located on the CoSSP template (see Fig. 4 and 5). The oil and gas stream will be commingled for transportation in a single pipeline to the production platform.

A conventional horizontal separator is proposed. The separator has been sized to allow a minimum water residence time of six minutes which should be sufficient to ensure a water quality of less than 1,000 ppm oil in water.

Water pumping will be performed by ESPs contained within a dedicated vertical caisson and operating in parallel. The 24" diameter caisson will be some 40 m long and installed into conventional 36" casing set into the seabed (for pumping system concept and configuration selection see section 6.5).

Important features of the proposed system are as follows:
- Robust conventional separation process.
- Use of "off the shelf" field proven pumping equipment.
- Maximum use of existing field proven subsea technologies.
- Use of ESPs in a pump caisson minimises leak paths to the environment.
- High system availability: use of parallel ESPs in shallow caissons allows ESP maintenance to be achieved at reduced cost. A redundant ESP can be installed in the caisson for little additional cost.
- Operational flexibility: Use of multiple pumps in parallel offers operational flexibility to meet changing production profiles without the need for a recycle loop. Use of "off-the-shelf" ESPs offer the possibility to choose the best pump and motor combination to meet changing field production profiles and varying injection pressures.
- Minimize life-cycle costs. The pump caisson has been designed for easy retrieval using a DSV. All critical instrumentation has been designed for diverless ROV installation and retrieval.
- Minimum subsea control complexity. All subsea control components are conventional and field-proven.

Control Philosophy

Water level in the separator will be measured by multiple redundant nucleonic sensors and controlled by varying the pump speed via the VSD units on the production platform. The control system will be designed to run without operator intervention. Operator alarms will be raised if a system malfunctioning is detected and the pumps will be shutdown automatically and the CoSSP system bypassed if the malfunction is not corrected within a set time period.

6.4 Pumping system

The pumping sub-system function is to boost the separated water pressure sufficiently for injection into the field aquifer at a design flow rate of 6,000 m3/d and wellhead injection pressure of 186 bar. Some flexibility in pumping capacity is required as water injection flow rates will vary from 1,000 to 4,500 m3/d in the first five years and it is not desirable to have a recycle loop. In later years, it will be desirable to inject more than 6,000 m3/d subject to the topside power limitation of 2 MW. The pumping system will be powered by VSDs located on the platform some 5 km remote from the CoSSP template. Pump speed (and thus flow rate) will be varied in order to control the water/oil interface level in the separator.

In the following the design basis for the pumping system are listed.

Pump design capacity (water only)	= 6,000 Sm3/d (38.000 s.bbl/d)
Suction temperature	= 50 - 65 C
Suction pressure	= 35 - 60 bara (500 - 870 psia)
Water injection pressure (max.)	= 190 bara (2690 psia)
Oil in water specification (max.)	= 1,000 ppm
Maximum topsides power	= 2 MW
Power supply voltage	= 11 kW
Power supply cable length	= 4,5 km

If water production exceeds injection pump then excess water will flow to the platform with the oil and gas production.

The following design philosophy has been adopted in selecting the proposed pumping concept:
- Use commercially available and field-proven components (pumps and motors).
- Minimize subsea mechanical, process and electrical connections.
- No subsea transformers or electrical switch gear.
- System should provide operational flexibility.
- No requirement for recycle loop.
- Positive pump NPSH required for long run life.
- Minimize offshore Inspection, Maintenance and Repair (IMR) costs (e.g. DSV installable).
- Design to minimize life-cycle costs.

6.5 Proposed pumping system configuration

<u>Pumping system concept and configuration selection</u>

Two generic pumping concepts were considered for the specific application:
- Marinized versions of conventional surface pump and motor sets.
- Electrical Submersible Pumps in caissons.

Preliminary analysis revealed that the ESP option had distinct advantages and could meet all the design philosophy requirements. The most significant attraction of the selected ESP in a caisson option was the fact that all components could be purchased "off the shelf" from more than one vendor e.g. Reda and Centrilift. The proposed pumps, motors, subsea electrical connectors and VSDs have been field-proven subsea and in similar water injection duties.

Various ESP and caisson configurations were considered in detail. E.g. multiple caisson versus a single caisson with multiple pumps. The evaluation also involved the analysis of variour combinations of parallel and series pumping configurations. The optimum configuration was found to be a single 24" x 40 m caisson with two (or possibly three) ESPs operating in parallel (see Fig. 6). A parallel pumping configuration was found to give the maximum controllability combined with operational flexibility.

Pumping system components

The proposed pumping system configuration is made up of the following components:

Pump caisson

The pump caisson is 24" diameter and some 40 m in length. The caisson contains two (or possibly three) ESP pump and motor sets. Each ESP pump and motor set is pre-installed into a 9-5/8" casing pipe suitably supported within the 24" caisson. The ESPs are hung-off from the caisson head assembly by the 4" pump discharge tubing and take suction from the bottom of the caisson. The process fluid flows up past the motor sets to provide cooling prior to entry into the pump suction. The discharge tubing contains a high integrity non-return valve. Note that the proposed ESP configuration is exactly the same as the standard ESP well configuration and no equipment modifications are required for caisson deployment.

Caisson head assembly

The caisson head assembly is connected to the caisson by a standard 24" flange which is made-up onshore at the integration site. The head assembly includes the following components:
- *Discharge Piping:* The 4" discharge pipe from each pump connects to a short 6" header pipe and collet connector. Each pump discharge pipe is fitted with an isolation valve which is ROV operable. The discharge piping and fittings are rated to 350 bar (5,000 psi).
- *Suction Piping:* The single 8" suction pipe and collet connectors are located on the caisson head (rather than the caisson shell) so that the head assembly can be utilized to test all wet mateable connectors during system integration without the necessity to have the caisson installed.
- *Electrical Connections:* Two standard Tronic (8 kV, 180 A) dry-mate penetrators are located in the caisson head and connect to a short jumper lead which terminates in a standard Tronic (8kV, 180 A) wet-mate connector which is made-up to the umbilical termination by ROV. The head assembly penetrators and internal caisson electrical connections are made-up in the dry and tested at the onshore integration site before shipment offshore.

6.6 Installation, Maintenance and Repair

Installation of large ESP sets in subsea wells is now a field-proven technology and the techniques for running the pump and motor strings into deep-water subsea wells through specially adapted Xmas-trees is also well developed. However the shallow depth of the pump caisson (e.g. 40 m) allows the well installation methods to be simplified significantly. Two simplified ESP installation concepts were carefully evaluated:

Offshore Make-up: Run ESP pump and motor strings into a caisson pre-installed on the subsea template.

Offshore Make-up: Pre-installed pump and motor sets in the caisson onshore and install the caisson as a single unit.

Onshore make-up was selected as the best option for the following reasons:
- Permits complete pump and motor integration test prior to offshore installation.
- Pumps and motors are protected by the caisson during installation.
- Minimizes subsea mechanical, process and electrical connections.
- Installation can be undertaken from a DSV.
- Redundant pump sets can be added to the caisson for little additional cost.

A typical installation sequence considering a DSV is shown in Fig. 7.

7   APPLICATIONS AND SIGNIFICANCE OF THE SUBJECT MATTER

"Subsea processing" is one of the challenging new technologies that can provide significant cost saving which help facilitate the economic exploitation of deepwater oil and gas reserves (expecially those that are marginals). Subsea processing improves field economics mainly by increasing production from the reservoir and/or reducing topside processing facilities.

Two main applications have been presented and discussed in this paper, namely: subsea separated water re-injection into the reservoir and raw water treatment and re-injection.

These two applications have been considered because of their significance in view of future fields development and both can be covered by the CoSSP concept with particular reference to be subsea water boosting and injection system.

The advances in the subsea separation technology can significantly contribute to both the increase of the well efficiency and the assurance of flow conditions from the wells to the surface receiving facilities. In fact, water removal from the well stream and re-injection into the reservoir can constitute an attractive option to reduce the total flowrate to be transported, to mitigate the effects of water on the production stream viscosity and the risk of hydrate formation and to avoid bottlenecks on the receiving treatment facilities.

The subsea raw water treatment and re-injection into the reservoir, to assure the reservoir pressure maintenance during the field life, has the benefit to increase the production and to reduce the topside receiving facilities.

The CoSSP concept and relevant building blocks provides Operators with a readily available system to face the requirements to use subsea processing as dictated by future deepwater developments. The above through a number of key features like: the maximum use of field proven equipment "off-the-shelf" availability of key components, minimization of the number of active component, minimization of the expected life-cycle costs by reducing the need and frequency of intervention as well as the requirement for small to medium size DSV to perform maintenance operations.

CoSSP system and the relevant subsystems has been designed to obtain operational flexibility integrating existing technologies and using field proven components; this lead to be able to supply an Operator tailor made system meeting the required schedule without any risk.

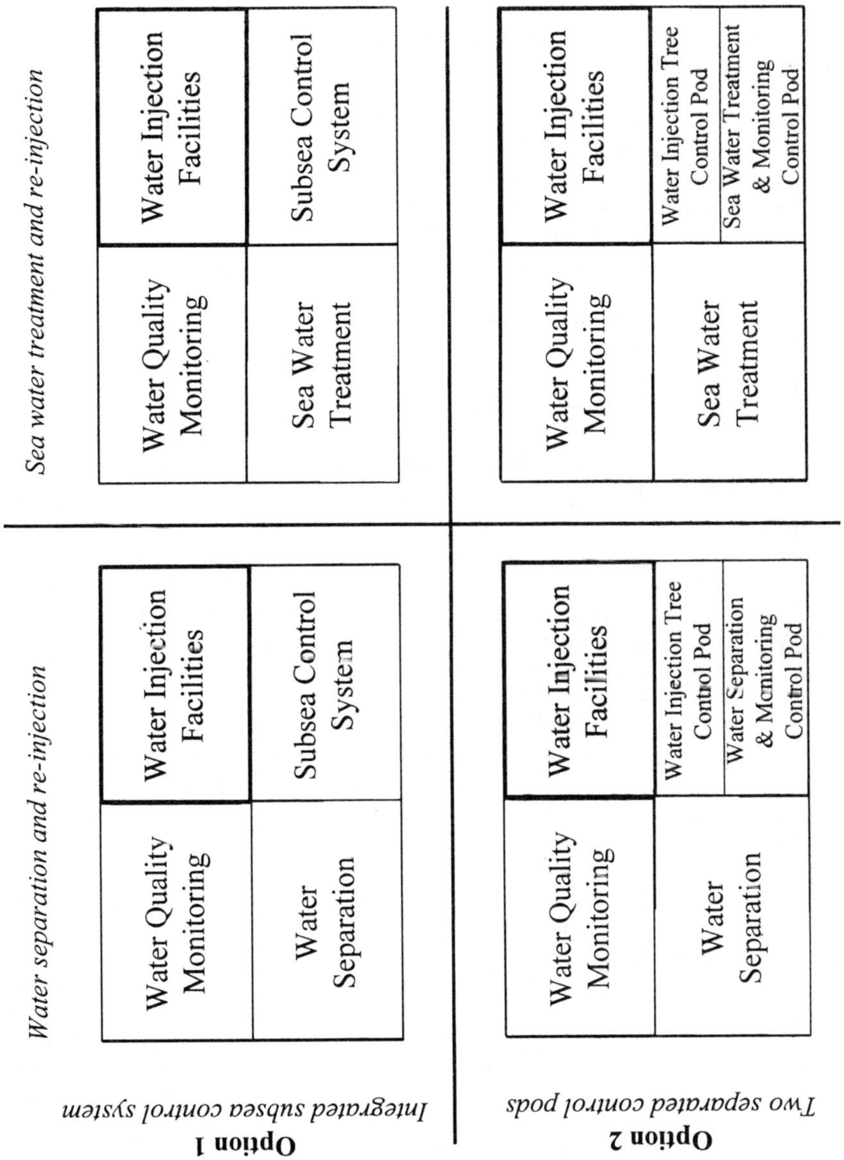

Water Quality Monitoring	Water Injection Facilities
Water Separation	Subsea Control System

Water Quality Monitoring	Water Injection Facilities
Sea Water Treatment	Subsea Control System

Water Quality Monitoring	Water Injection Facilities	Water Injection Tree Control Pod
Water Separation		Water Separation & Monitoring Control Pod

Water Quality Monitoring	Water Injection Facilities	Water Injection Tree Control Pod
Sea Water Treatment		Sea Water Treatment & Monitoring Control Pod

Option 1
Integrated subsea control system

Option 2
Two separated control pods

Fig. 1 - Main subsea architecture options considering integrated configurations

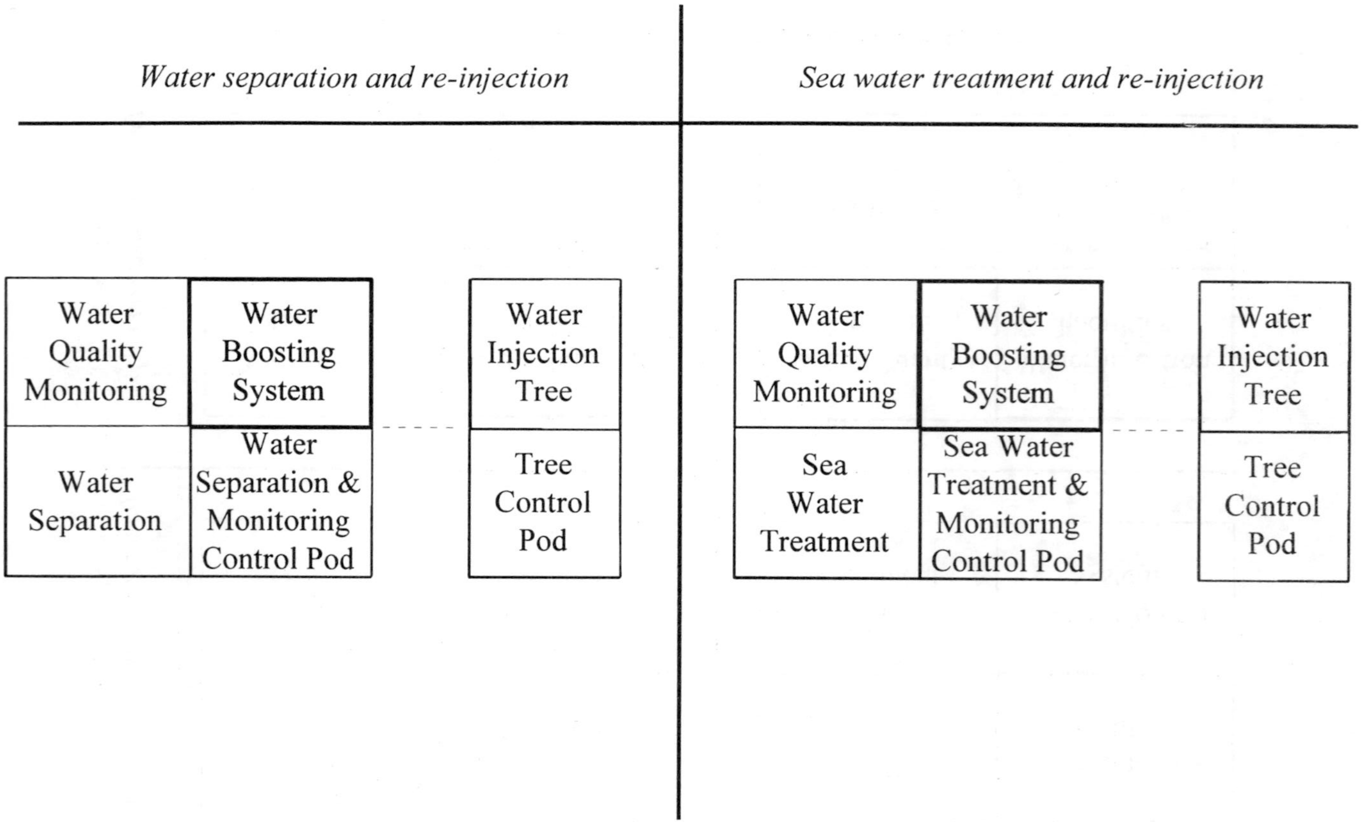

Water separation and re-injection

Sea water treatment and re-injection

| Water Quality Monitoring | Water Boosting System |
| Water Separation | Water Separation & Monitoring Control Pod |

| Water Injection Tree |
| Tree Control Pod |

| Water Quality Monitoring | Water Boosting System |
| Sea Water Treatment | Sea Water Treatment & Monitoring Control Pod |

| Water Injection Tree |
| Tree Control Pod |

Fig. 2 - Subsea architecture considering a distributed configuration

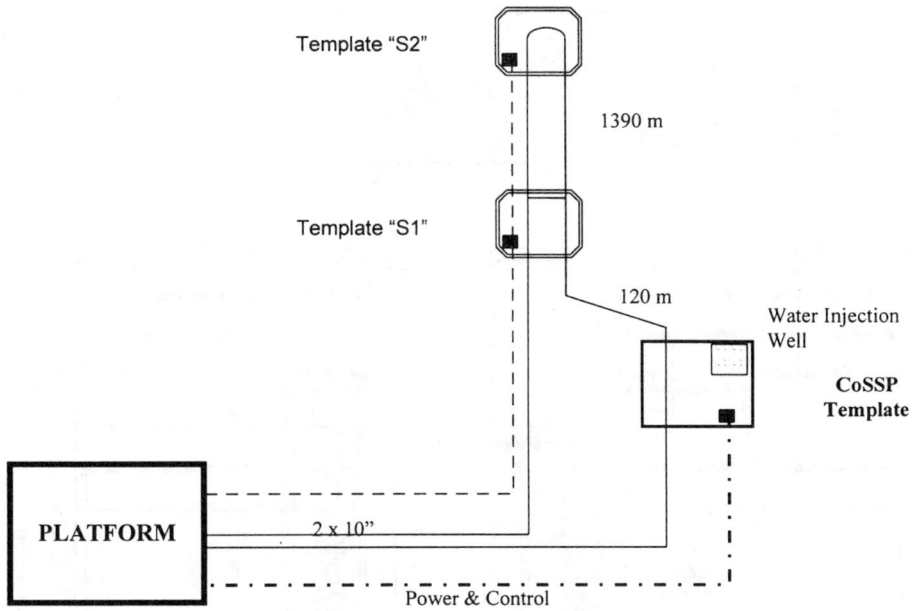

Fig. 3 - CoSSP template location

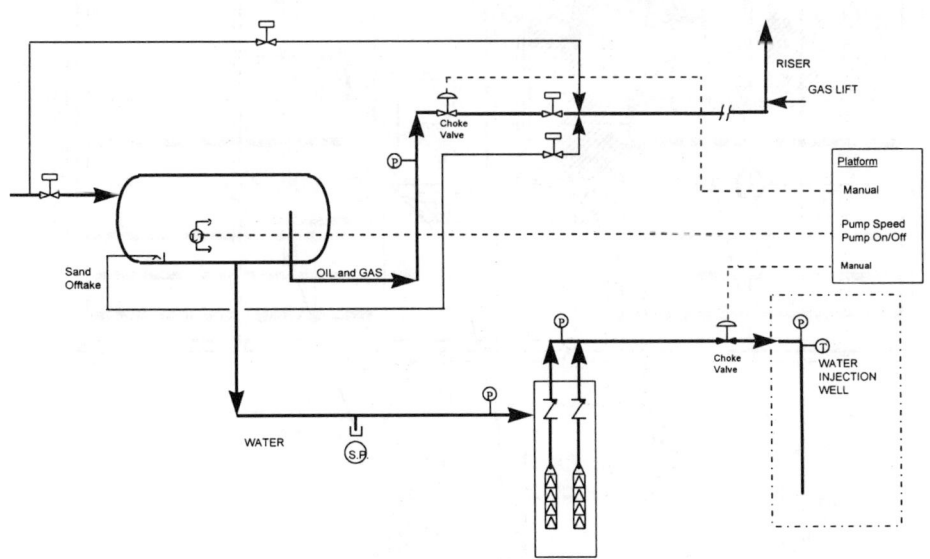

Fig. 4 - CoSSP Process flow diagram

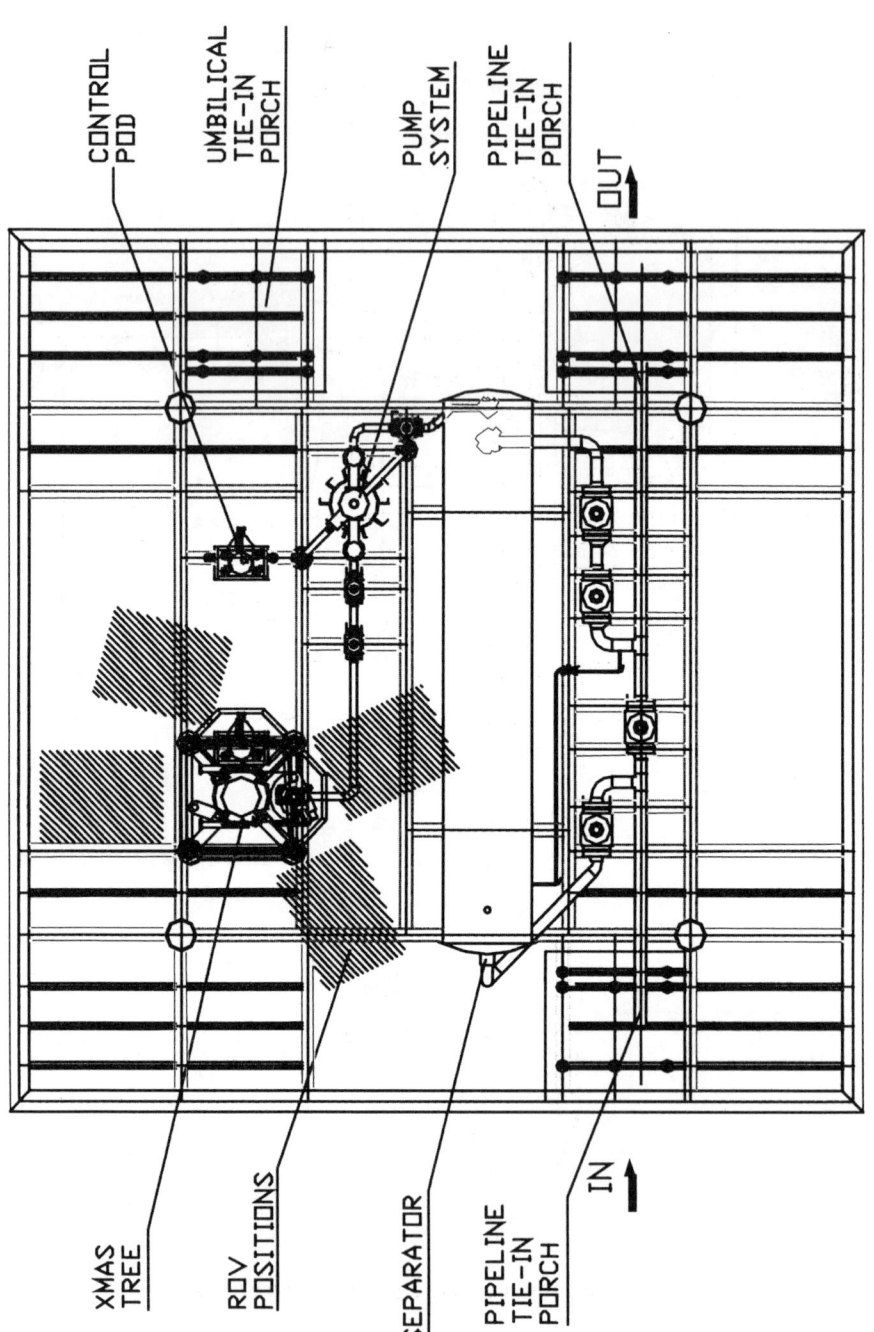

CONTROL POD

UMBILICAL TIE-IN PORCH

PUMP SYSTEM

PIPELINE TIE-IN PORCH

OUT

XMAS TREE

ROV POSITIONS

SEPARATOR

PIPELINE TIE-IN PORCH

IN

Fig. 5 - CoSSP general arrangement

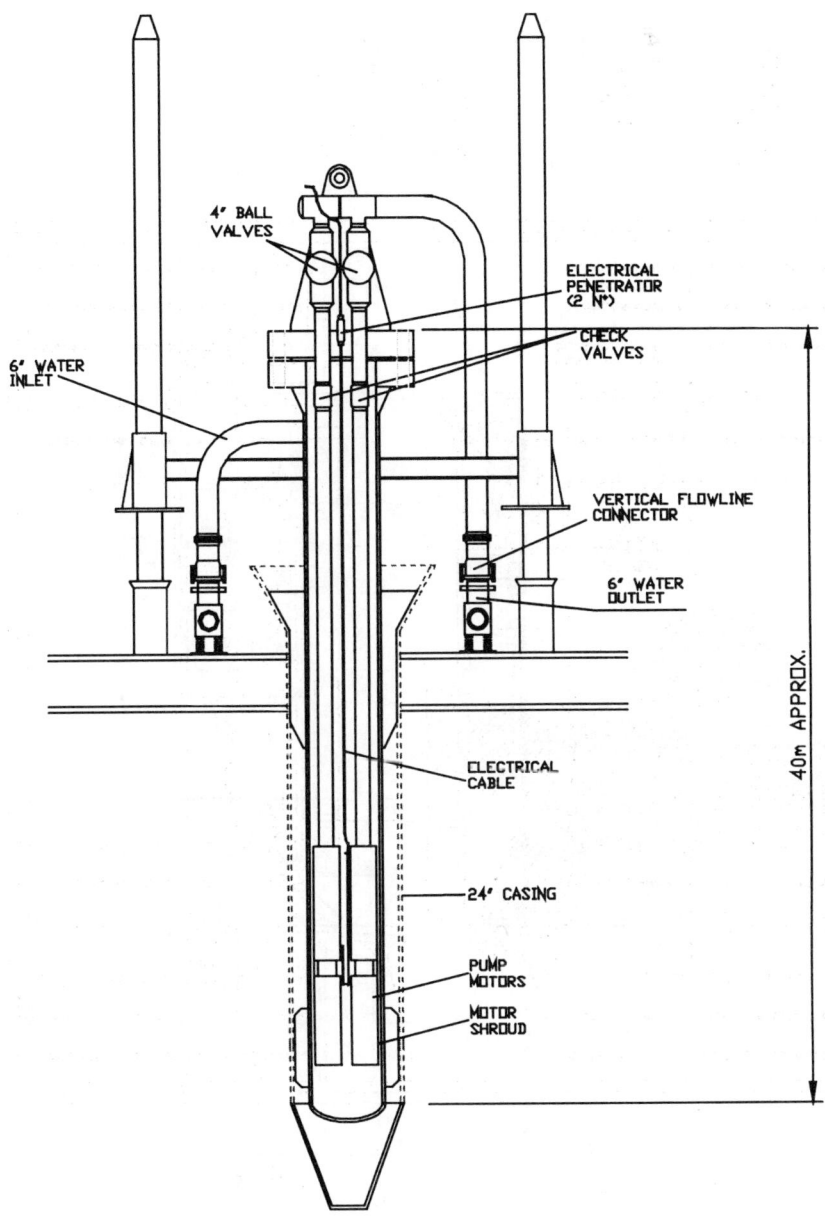

Fig. 6 - CoSSP pumping system general arrangement

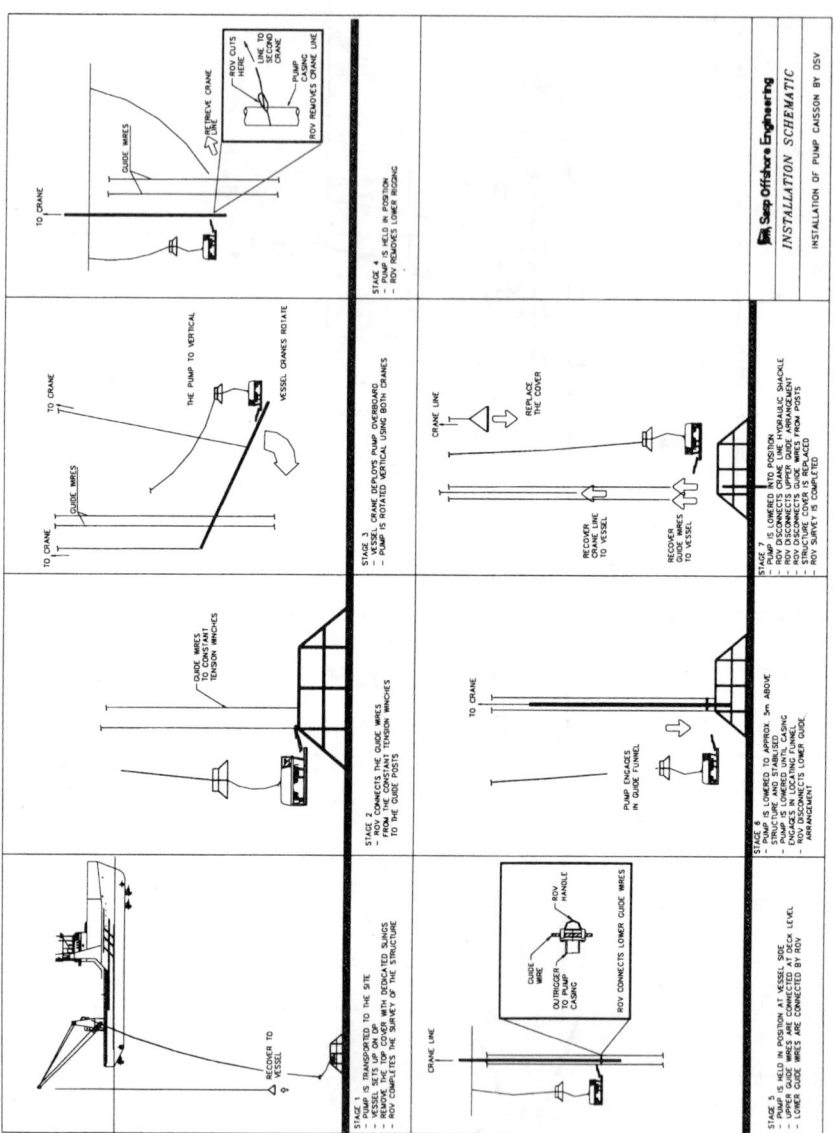

Fig. 7 - CoSSP Pump Caisson Installation Sequence

© BHR Group 1998 *Downhole Production*

Process aspects of the Troll Pilot Subsea Separation and Reinjection System

B BRINGEDAL and **K HAUGEN**
ABB Offshore Technology
M N LINGELEM and **B STRAND**
Norsk Hydro Technology and Projects, Norway

1. INTRODUCTION

Subsea separation has been considered for years, but only recently has the industry gained sufficient interest and confidence to it to look for possible applications. Possible reasons why such a long time has gone to mature the idea of subsea processing are:

- Conservatism and doubt to whether satisfactory availability can be achieved.
- Water treatment for disposal to sea is considered too complicated for a subsea application.
- Little experience with reinjection of oily water with respect to reservoir plugging, reduced injectivity and need for fracturing.
- Reinjection of produced water requires higher investment and operating cost compared to disposal to sea. Subsea separation and reinjection may therefore not be cost efficient if pressure support is not needed.

A major uncertainty with subsea water removal and reinjection, is the availability of the system. The benefits may be eliminated by lower availability caused by operating problems by introducing a more complex system subsea.

To gain operating experience and hopefully increase production with lower water flowrates to topside, the Troll license group; Statoil (74.6%), Norsk Shell (6.3%), Norsk Hydro (7.7%), Saga Petroleum (4.1%), Elf Norge (2.4%), Conoco Norge (2.0%) and Total Marine Norsk (1.0 %) with Norsk Hydro as operator, initiated a project (start early summer 1997) to install a Pilot Subsea Separator and Water Reinjection System on the Troll Oil Gas Province (TOGP). The system which is called Troll Pilot shall be installed on the Troll Field during 1999.

ABB Offshore Technology in cooperation with Framo Engineering and ABB Corporate Research who in late 1995 had been working on a development project, SUBSIS for **Sub**sea Separation and **I**njection **S**ystem, had together gained sufficient competence in the area to win the contract for design and development of the Troll Pilot. wptih ABB Offshore Technology has been the main contractor with the following ABB subsuppliers:

- ABB Seatec – Subsea Control System
- ABB Industri – Power systems
- ABB Corporate Research - Assistance to ABB Offshore Technology for development of process system design, subsea power connector and inductive level detector.

The main external suppliers are:
- Framo Engineering – Water injection pump and variable speed motor
- Norsk Subsea Cable – Umbilical
- Proser – Separator Internals
- Motherwell Bridge – Separator pressure vessel
- Spectris Norge/BTG - Nucleonic Level detector

The objective with this paper is to give a brief discussion of potential benefits in general for a subsea separation and water reinjection system, then describe the Troll Pilot process system design with focus on system operation and less focus on detailed design.

2. DISCUSSION OF POTENTIAL BENEFITS WITH SUBSEA SEPARATION

Below is a brief discussion of the potential technical, economical and environmental advantages and disadvantages of subsea separation in general, and of Troll Pilot in particular.

2.1 Tie-in distance and water depth

Probably the most attracting potential advantage with SUBSIS is the potential of increased tie-in distances and sea depths; i.e. the fluid becomes transportable over longer distances and deeper sea within economical constraints. This may be possible due to removal of water from the flowline, which again gives the following benefits:

- Problems with hydrate formation may be fully alleviated or can be avoided with substantially cheaper solution, e.g. no thermal insulation and no or little use of hydrate inhibitors. This is further discussed in chapter 2.5.
- Carbon steel can be used since corrosion rates are low or can be controlled by use of corrosion inhibitors.
- For low-pressure reservoirs, the pressure may be insufficient to overcome the pressure drop in the system. However if water is removed subsea as with SUBSIS, the frictional and static pressure drop is reduced. The maximum flowline length and riser height can then be increased correspondingly. An alternative to subsea separation is increased number of flowlines and risers, but at large distances and sea depths the cost will be higher than SUBSIS. In addition, increasing the number of risers may even result in increased pressure drop, because gas-liquid slip velocity and hence liquid holdup and static pressure drop increases with reducing velocity.

However, SUBSIS also have some disadvantages with respect to increased tie-in distance. Assuming that the oil production rate, the thermal insulation and the number and size of flowlines/risers are the same with and without SUBSIS, the temperature drop increases when water is removed subsea. This gives the following potential disadvantages:

- Hydrate and wax formation temperatures are reached after shorter distances along the flowline with than without SUBSIS.
- Significant wax deposition may start in the flowline when water is removed subsea. This is further explained in chapter 2.5.

2.2 Investment costs

For a given production profile from the reservoir, subsea water removal may reduce the investment costs by allowing reduced number or size of flowlines/risers and/or size of topside equipment (e.g. separators, cyclones, crude heaters). Alternatively, with the same production/processing facilities except SUBSIS, water removal subsea may allow increased production rates as discussed in chapter 2.4.

Other potential investment cost reduction elements due to SUBSIS are:

- Smaller topside slug catchers and separators as discussed in chapter 2.6.
- Smaller or no subsea depressurization vessels due to reduced amounts of liquid to topside during subsea depressurization. The reasons are lower liquid holdup when water is removed, and that depressurization for hydrate inhibition may not be required if the amount of water is sufficiently low that hydrate formation can be disregarded.
- For concepts with flowline directly to shore, the cost of entire platforms can be saved.

These benefits obviously need to be evaluated against the cost associated with SUBSIS and compared to alternative solutions like increased flowline dimension, increased topside capacity etc. The economical benefit of SUBSIS will therefore be field specific. Need for pressure support in the reservoir, longer flowlines, deeper water and higher water production are parameters which will tend to favor SUBSIS.

2.3 Operating costs

As mentioned in the introduction, a major drawback with SUBSIS may be low availability due to the increased complexity for the subsea system and production problems due to sand, wax, hydrates, scaling, emulsifying crude etc. These problems will increase the operating cost compared to topside processing by:

- Time consuming subsea interventions, limited access to solve the problem and less information available. This will result in delayed production and for some cases reduced oil recovery, and increased operating cost per barrel oil produced.
- Increased repair cost.

Troll Pilot will hopefully show that these are not as significant problems as many fears, and instead prove that SUBSIS will reduce operating costs from the following reasons:

- Reduced usage of production chemicals due to less water in the flowline as discussed in chapter 2.4.
- Reduced pump power for water injection, comparing SUBSIS with topside reinjection of produced water, due to higher separator pressure for SUBSIS than for topside, i.e. lower pressure head. However it may also be opposite, because of the following reasons:
  - Higher injection pressure due to more oil and solids in the water from a subsea separator than from a topside system.
  - Due to the potential of hydrate formation in the injected water the temperature must be above the hydrate formation temperature. Compared to injection with cold water, thermal fracturing will be reduced which will increase the power consumption.
  - Injection topside gives an increased water column above the water injection well. This water column represents a higher head than the static pressure drop through the production well, flowline and riser due to natural gas-lift up the well and riser.
- Injection of seawater may give severe sulfate scaling and formation of H_2S in the reservoir. These problems are avoided if produced water is injected instead.
- Reduced topside compression power due to less flowline/riser pressure drop and hence higher topside separation pressure.
- For directly to shore concepts, one saves cost of the offshore processing.
- Improved separation properties may reduce operating cost by less downtime, less use of separation improving chemicals and reduced heating requirements, see chapter 2.5 below.
- Reduced water slugs and possibly also reduced slugging tendency in general will reduce operating difficulties and thereby also the cost. Increased slugging tendency may on the other hand have a correspondingly negative effect, see chapter 2.5 for further discussion.
- Possibly easier start-up of drowned wells. The reason is such start-up (kickoff) requires low wellhead pressure and high liquid flowrate that may not be possible without water removal close to the wellhead(s).
- More rapid start-up since the wellhead pressure can be reduced and/or tedious hydrate prevention procedures may be simplified.
- The subsea separator can in some cases be used as a test separator, using reliable measurements, which may be advantageous due to shorter time to establish steady state than topside with long transport distance from well to test separator.

2.4 Oil recovery and production rate

SUBSIS may increase production rates because water removal subsea may allow increased production rates. This will then give faster reservoir drainage and increased net present value (NPV) of the oil. This benefit is an alternative to reduced investement costs and it assumes that the oil production rate is limited by subsea or topside fluid handling capacity.

The NPV may obviously also be increased by increasing the oil recovery. For low pressure reservoirs with no natural drive or no water/gas injection for pressure support (as Troll Oil), one may get increased recovery when water is removed at the seabed because:

- The wellhead pressure may be reduced while still maintaining production to the platform.
- Production with higher watercuts may become technically feasible because removing the water may reduce the pressure drop such that the reservoir pressure is sufficient to lift the fluid through the system and up to the platform.
- Production with higher watercuts may become economically feasible because of increased well production rates and/or decreased capital costs (lower investment).

2.5 Consumption of production chemicals

Use of the following chemicals may be reduced or avoided due to subsea removal of water:
- Scale inhibitors. The reason is that water removal may cause a transition from water continuous emulsion with a water film at the wall to an oil continuous emulsion with an oil film at the wall. With no water film at the pipe walls, scaling should be minor or zero.
- Corrosion inhibitors. The reason is that corrosion is low with oil wetted pipe walls.
- Hydrate inhibitors. The reason is that with sufficiently little water in the flowline, hydrate formation will be is a minor or nonexisting problem. (However as discussed in chapter 2.1, the removal of water may reduce the temperature in the flowline which may increase the need for hydrate inhibitors).

The transition from water to oil wetted pipe walls may create a serious wax problem such that wax inhibitors may be more needed with than without SUBSIS. Another negative effect with removal of water is reduced temperature, assuming same thermal insulation and oil flowrate.

2.6 Separation and slugging aspects

The following separation benefits may exist for and/or due to SUBSIS:
- Upstream slug sizes are expected to be significantly smaller than for topside separation since severe riser base terrain induced slugging will not occur in a subsea separator.
- Reduced water slug sizes into the topside processing facilities.
- Installed at the riser base it may act as a slug catcher or avoid riser based slugging. The gas and liquid may be transported up in individual riser. If the liquid is pumped up, one may either allow lower seabed pressure or reduce topside compression and total power.
- More stable flowrate into a subsea separator than into a topside separator, regardless of slugging, because the upstream surge volume generally will be significantly smaller for a subsea separator than for a topside separator.
- Possibly more easy separation subsea than topside due to higher pressure and temperature and hence lower liquid and emulsion viscosities, larger oil-water density difference and finally more fresh droplet surfaces (which gives less stabilizing of emulsions due to diffusion of surface active components to the droplet surfaces).
- Compared to floaters one avoids motions.
- Possible lower consumption of methanol may reduce processing problems in downstream systems (separation and refinery systems).

The following topside separation disadvantages may be due to SUBSIS:
- Lower temperature to the platform or increased insulation thickness. This may result in poorer separation in the first stage separator. However, the combined efficiency of the first stage and subsea separator is likely to be better than for only a fist stage separator only (even if the total volume is equal for the two cases).
- Increased slugging tendency due to lower flowrates if SUBSIS is used to reduce the pressure loss. This is not relevant if SUBSIS is used to reduce flowline dimensions.
- Larger fluctuations inlet conditions to topside due to variation in SUBSIS water injection flowrate and/or subsea separator level.

2.7 Environmental aspects

The main environmental benefits with SUBSIS are related to reduced disposal of produced water to the sea. The produced water contains production chemicals, heavy metals and sometimes radioactive components in addition to dispersed and dissolved oil in water.

Additional potential environmental benefits are:
- Reduced usage of production chemicals as discussed in chapter 2.5.
- Less flaring due to less need for depressurization for hydrate prevention and less production upsets due to slugging (this may however go both ways).
- Reduced compression power (i.e. less CO2 to the atmosphere). By separating gas and liquid subsea and by letting gas and liquid flow in separate lines to topside, the demand for topside recompression can be greatly reduced. In this manner, subsea separation technology can contribute to a better utilization of resources and to a reduced environmental load.
- Increased recovery. This is environmentally beneficial because new field developments and manufacturing is delayed.

Regarding leaks and risk of ruptures with oil spill, there are two opposing factors:

- More complex subsea system increases the risk of leaks and spills.
- Possibly less flowlines reduces the risk of leaks and spills.

Potential environmental disadvantage with SUBSIS is increased power consumption (and more CO_2 to the atmosphere), comparing SUBSIS with either the dispose to sea alternative, or with the clean cold water injection from topside.

ABB has ongoing work with Life cycle analyses of typical SUBIS concepts. Preliminary conclusions indicate that SUBSIS gives reduced overall environmental impact compared to conventional topside installations, mainly due to reduced energy consumption.

2.8 Troll Pilot

Oil Production from the Troll field poses significant operational challenges due to the thin oil layers such as a large span of water rates, gas to liquid ratios and subsea separator pressures. Hence, if success of Troll Pilot is demonstrated, the creditability of such systems should be high enough that other fields may also utilize this technology.

The potential benefits for the Troll Oil production are as mentioned in chapter 2.1 – 2.7.
The Troll Field is especially well suited for the application because of the following reasons:

- The reservoir pressure is low (dropping over time from initially hydrostatic pressure) such that reduced wellhead pressure may give increased oil production and considerably less risk of "drowned" wells.
- Start-up (kickoff) of drowned wells requires low wellhead pressure and high liquid flowrate that may not be possible without subsea water removal.
- Very high expected water production from Troll oil. Hence reinjection of produced water will significantly reduce the environmental load compared to water disposed to sea.
- The flexibility to handle an unforeseen production increase will be improved.
- Optimized production strategy for the thin oil layers as on Troll, by moving the horizontal well sections deeper/closer to the WOC (water oil contact) and further away from the gas column. This will reduce the gas production and potentially increase the oil production and recovery.

However, since the design of Troll C is made such that the production is expected to be limited by gas instead of water processing capacity, seabed separation may not give any significant positive economical contribution for Troll C.

Regarding the environmental arguments, these are not yet used on Troll as it is still uncertain whether reinjection of produced water gives less environmental impact than dispose to sea.

For future field developments the benefits may however be fully utilized if the operation of Troll Pilot proves to be satisfactory.

3. PROCESS SYSTEM DESCRIPTION

3.1 Total system

The Troll Pilot subsea separation system is designed to remove and reinject bulk quantities of water from the wellstream of one production line from two well clusters as shown in figure 1.

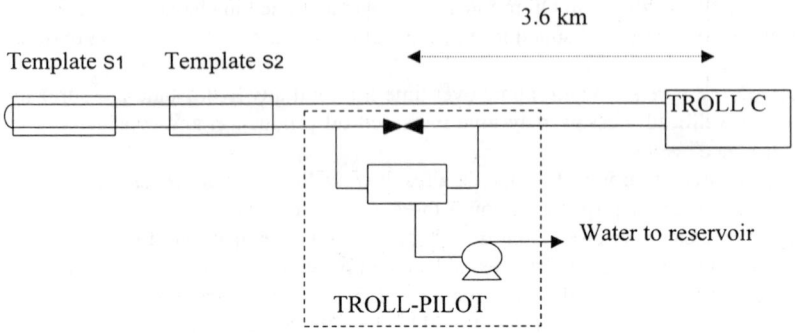

Figure 1. Troll C Subsea System for S1, S2 and Troll Pilot

The wellstream enters a horizontal gravity separator where free water is removed. Oil and gas are re-mixed at the separator outlet and transported through the production line to Troll C. The separated water is boosted by the re-injection pump and injected into the reservoir through a dedicated injection well. Injection without the water injection pump may also be possible.

The design of Troll Pilot is based on the following design capacities:

Maximum oil:	6 000 Sm³/D
Maximum water:	9 000 Sm³/D
Maximum liquid:	10 000 Sm³/D
Maximum gas:	800 000 Sm³/D at 60 bara

A simplified PID of the system is shown in figure 2.

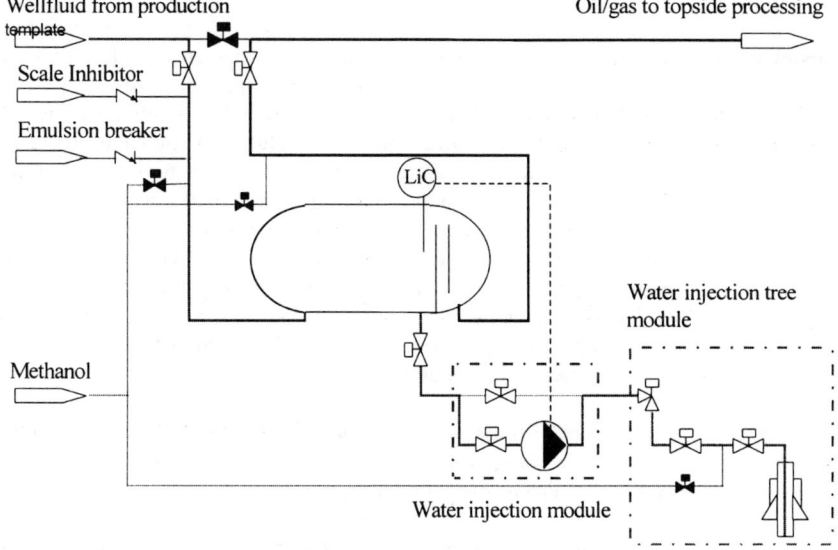

Figure 2. Troll Pilot process system

Stable separator oil and water levels are required in order to obtain stable conditions and required retention times for the separation process as well as stable flow towards Troll C. This will be achieved by closed loop control of the water level by controlling the pump speed and by control of the oil level by the outlet baffle arrangement. Two independent and functionally different level detectors monitor the water level. In addition, oil level and emulsion layer thickness may be monitored.

Measurement of the water quality is achieved by means of an ROV operated sample point located at the re-injection pump inlet whereas oil quality can be measured at Troll C.

The system is equipped with the following by-pass possibilities:

- By-pass of the entire Troll Pilot.
- By pass of the water injection system by closing the separator water outlet. This is the normal bypass mode in case shutdown of Troll Pilot.
- By pass of the water injection pump for reinjection without boosting the injection water.

3.2   Separator

The separator is a conventional horizontal separator. The internal diameter is 2.8 m and the length (tan-tan) 11 m. The main reasons for choosing a horizontal separator were:

- To limit the uncertainty with respect to separation of water and oil.
- Experience with the Troll fluid indicates a difficult separable fluid due to high oil density and tight emulsion characteristics. Build-up of a free water phase will typically require a long retention time, which is most easily obtained in a horizontal separator
- Sand production is expected to be minimal during normal production. Most of the sand is expected to be of small size, e.g. fines which will not accumulate in the separator but follow the fluid flow. The separator is by-passed during periods were sand production may be expected, for example during start-up of new wells. Additional protection is achieved by means of an intervention based jetting system.

A surge chamber to ensure oil supply to the Troll C production line without oil supply to the separator for a limited period is incorporated in the design. The purpose is to avoid introduction of flow instabilities in the downstream piping.

A new inlet device is developed in cooperation between Norsk Hydro Technology & Projects, and ABB Offshore Technology. The main principles are separation of gas/liquid before the liquid is exposed to high shear, and minimization of shear by smooth reduction of momentum. Based on successful test results with the new inlet, it was confirmed that coalescing internals were unnecessary. This may be very beneficial to avoid plugging by sand/asphaltenes/scale.

3.3   Water Injection pump

The water injection pump is a rotodynamic helico-axial pump based on the Poseidon technology and represents an extrapolation of design previously developed and implemented in various deliveries. It consists of two main components:

- The barrel which is permanently bolted to the water injection module.
- The retrievable insert containing pump impeller, diffusor and motor.

A variable speed motor drives the pump. Maximum power to the motor is 2 MW from topside Troll C and maximum speed is 3600 rpm.

3.4   Injection tree

The water injection tree is based on a standard TOGP production tree with some minor modifications. The tree includes a remotely controlled production choke that can be used for:
- Improved stability of the closed loop control as described in chapter 4.4
- Simplified water injection start-up as described in chapter 4.5
- Separator water level control in periods when injection can be achieved without the pump.
- Increased required pump head and hence speed if pump speed drops below minimum speed of 1000 rpm.

3.5   Chemical injection systems

The following chemicals and control fluids are supplied through a control and service umbilical from the platform:

- Scale inhibitor for prevention of calcium carbonate deposits.
- Emulsion breaker to break stable water in oil emulsions.
- Methanol for hydrate inhibition and removal. Three injection points are provided; at separator inlet pipe, at oil outlet pipe, and at the water injection Xmas tree. The methanol system is also used for pressure equalization before opening of valves.
- Hydraulic oil based control fluid for opening/closing of remotely operated valves.
- Lip seal fluid to the WI pump for provision of sufficient pressure to seal off the insert from sea and to seal off the different internal pressure levels.
- Barrier fluid to the WI-pump which forms a seal to stop injection water entering the motor, as well as serve as lubrication for the pump, cooling medium for the motor, dielectric media for the motor and corrosion protection of motor and pump internals.

3.6   Sampling system

A ROV operated sampling system has been provided for measurements of the water quality from the separator. The sample point is located on the water injection module upstream the WI pump by-pass line.

4. SYSTEM OPERATION

4.1 Operational strategy

The main operational strategy is to inject as much water as possible within the given constraint; available power (2 MW from platform), minimum allowed arrival temperature to Troll C and maximum 1000 ppm oil in water.

To achieve this the separation efficiency should be optimized during normal production, and the number and duration of shutdowns should be minimized.

Due to limitations in the power supply at Troll C the water injection rate will be less than the water production rate in the late production phase. The injection capacity will depend on the required injection pressure and is approximately 6000 m^3/d for the maximum fracturing pressure. Excess water will follow the oil/gas mixture to the platform if the capacity is exceeded. Hence there is no strict requirement on water carryover to the oil outlet. However, if the water rate to Troll Pilot is less than capacity water injection capacity, maximum 10% water in oil shall be achieved.

4.2 Main operational requirements

- The operational procedures for Troll Pilot shall be consistent with procedures for the overall subsea system. Hence the separator and associated manifold piping is considered part of the flowline during any shutdown or startup. During shutdown of Troll Pilot only, production is routed through the separator.
- It shall be possible to shut down water injection and/or isolate Troll Pilot without stopping production from any wells.
- Before opening of all valves, except needle and choke valves, pressure equalization shall always be performed.

4.3 Hydrate prevention

The following requirements with respect to hydrate prevention are given:

- Injection water from the subsea separator shall be treated as a hydrate forming fluid, because the water contains dissolved gas and small quantities of oil droplets.
- Methanol shall be used as hydrate inhibitor fluid.
- It shall be possible to inhibit all parts of Troll Pilot with methanol.
- It shall be possible to depressurize Troll Pilot at the same rate as the subsea flowlines.
- A decision time of 4 hours shall be allowed before any hydrate prevention is needed.
- With normal liquid level in the separator a cooldown period of 18 hours shall be allowed without need for hydrate prevention actions.

The methanol injection capacity is approximately 5 m^3/h. This is more than sufficient to fulfil the above requirements.

Hydrate prevention during the various operational modes is achieved as follows:

1) Normal operation (production and injection) and short term shutdown
All parts of Troll Pilot will be thermally insulated to maintain inner wall temperatures above the hydrate formation temperature during normal production and short term shutdowns. Therefore no hydrate prevention actions are required during normal operation.

2) Long term shutdown of water injection
If only water injection is shut down, production is maintained through the separator in order to avoid temperatures below the hydrate formation temperature in the separator system. Thus, only the water injection system is inhibited with methanol.

3) Long term shut down of well production to Troll Pilot
Hydrate prevention will be achieved by depressurization to Troll C together with the flowline.

4) Long term shut down of production through Troll Pilot (e.g. by isolation)
In this case the separator and the associated piping must be inhibited with methanol.

5) Cold start-up of Troll Pilot together with cold flowline
Circulation with hot stabilized oil through the separator is a possible method. This requires no special action from Troll Pilot. When circulation is finished, water injection start-up can be made just as after short-term shutdown, except that the water in the WI-system is displaced with methanol.

However there are uncertainties with respect to:
• Sand accumulation into the separator outlet during circulation in reversed direction
• Time required to heat the water with the oil flowing on top of the water.

Therefore it is recommended to avoid hot oil circulation through the separator, and instead inhibit the water in the separator with methanol prior to start-up. Hence start-up is performed as for the "cold start-up after isolation of Troll Pilot" as described below.

6) Cold start-up after isolation of Troll Pilot
Methanol injected during shutdown will prevent hydrate formation in the separator during flowline start-up. Water injection start-up can then be made as after a short-term shutdown, except that the WI-system is displaced with methanol.

4.4 Dynamic performance

The dynamic stability of the process system has been given thorough attention due to:
• No pressure control.
• Potential slug generation from the combined oil and gas outlet.
• Multiphase flow operating in the slug regime both up- and down-stream Troll Pilot.
• Large operating range of pump as well as potential operation in surge regime.
• No experience with closed loop control subsea.
• Interaction with the water injection well.

Extensive simulation activities by using state of the art simulation tools have therefore been an important activity in the design work. The system has been simulated by using the D-SPICE dynamic process simulator linked to the "OLGA multiphase pipeline simulator" and the system stability was tested with different control algorithms for a wide range of operating conditions as well as internal and external disturbances.

4.5   Water injection system flow characteristics

The water injection system flow characteristics is here defined as follows:
The curve described by the pressure at the pump discharge (upstream the injection choke) as function of the water injection flowrate. For the pump this curve is called the pump curve or pump characteristic. For the injection well including the choke and the inflow to the reservoir the curve it is called the system curve or system characteristic.
The system and pump curves are shown in figure 3.

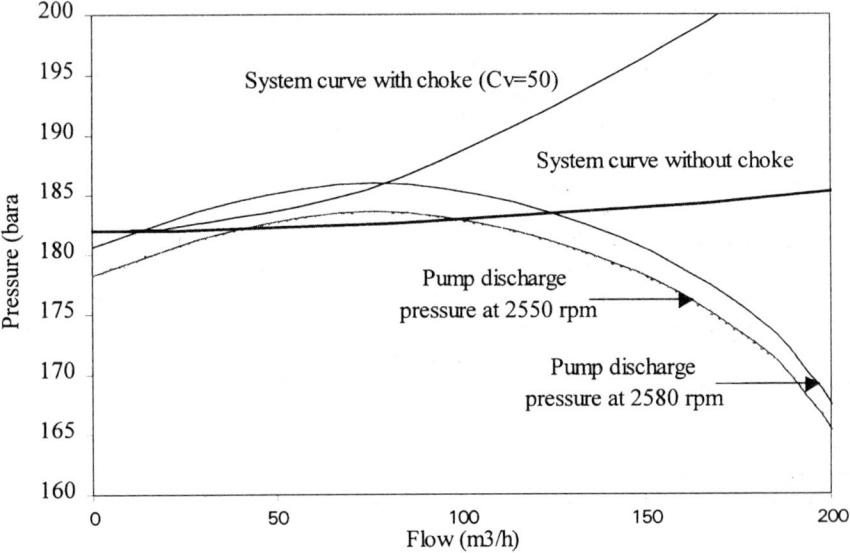

Figure 3. "Pump curves" and "system curves" with and without choking.

The system characteristic is highly uncertain because no experience exists with injection of produced water into a highly unconsolidated reservoir as on Troll Oil. It is conservatively assumed that the reservoir will plug up by sand and oil (which includes asphaltenes and resins) and that it is necessary to constantly fracture the reservoir to maintain injection. Hence the required pump discharge pressure is equal to the fracturing pressure (minus the static head in the well). Without a choke pressure drop in the system, the system curve becomes close to "horizontal" as shown in figure 3, because injection is performed through the casing such that the frictional pressure drop is low. The slight inclination is due to friction in the short piping and tubing above the downhole safety valve. By reducing the choke opening one is however able to tilt the system curve.

This may be useful because the water injection pump curves, as indicated in figure 3 for two typical pump speeds, is expected to give a maximum head point (although it needs verifications with pump performance tests). Hence the pump gives the same head with two different flowrates and both combinations of flowrate/pressure are possible, because the pump and system curve crosses in two points. The result is that external disturbances may induce unfavorable switches forth and back between the two operating points.

As seen in figure 3, introducing a choke pressure drop may alter the system characteristics such that only one flowrate is possible for the same injection pressure.

4.6 Water Injection start-up

Since reservoir behavior during start-up is uncertain it is conservatively assumed that fracturing is required before flow starts into the reservoir. The well/reservoir inflow behaves in principle as a closed valve that opens at the fracturing pressure. In this situation it is necessary to start the pump with circulation through the pump bypass. Without circulation, the pump will be operated below the minimum flowrate as explained below:

The injection pump will operate against a high resistance before fracturing is achieved. Depending on the degree of resistance the pump will either work on the "left" or "right" side of the maximum head point on the pump curve. Worst case will be closed reservoir, which may result in operation on the left side. This is very unfavorable because of heavy internal recirculation which gives vibrations and rapid temperature rise. In addition it will create problems to increase the flow to above the maximum head point as explained below:

Assume that start-up is performed with fully open production choke such that the system curve is as without a choke in figure 3. Further assume that the pump speed is 2550 rpm and that the pump curve crosses the system curve to the left of the maximum head point. Increasing the pump speed then moves the crossing between the two curves to the left, i.e towards reduced flowrate.

Operation to the left of the maximum head point could be feasible if the vibrations and temperature problems are acceptable with respect to pump wear, but it will be difficult to control due to the opposite relation between speed and flow.

To avoid this problem, circulation is performed until the pump speed has reached minimum 2550 rpm (for this specific case). Since there is very little pressure drop in the circulation loop, the flowrate will be far above the one that gives maximum head. When the bypass/circulation valve is closed, the crossing between pump and system curves will then occur on the right side of the maximum head point.

The main principles are therefore:

- Start injection through the pump bypass while the pump suction is closed. Then water will flow into the well by itself. If the water level drops, pump start-up is not required and the water level can be controlled by the remote operated choke.
- If not, open pump suction, start the pump immediately after to avoid forced rotation of the pump. Now some water will be circulated through the pump bypass and some (or no) water is injected. This ensures that the flowrate through the WI-pump is above the "min-flow" during and after start-up. See figure 3 and the explanation below.
- As fast as possible increase pump speed to required operating speed (predefined on basis of calculations and operating experience).
- When this speed has been reached, the WI pump bypass valve is closed immediately. Rapid closure is required to avoid excessive temperature rise.

Since zero flow into the reservoir may occur until sufficient speed and discharge pressure from the pump has been achieved, there may be significant temperature rise during start-up.

The figure below shows the pump speed and water temperature versus time.

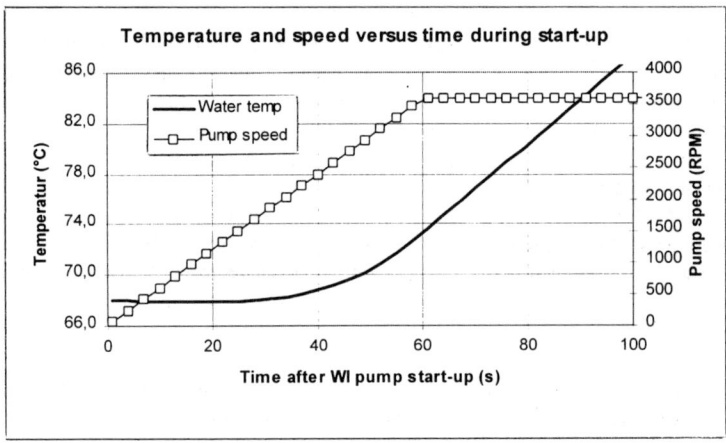

Figure 4. Water injection pump speed and water temperature versus time.

It is assumed that the fracturing process is rapid (due to unconsolidated and highly permeable sand reservoir).

A possible simplification to the base case startup procedure is use the injection choke such that circulation around the pump is avoided. If the choke opening is reduced to give a system curve as for example shown on figure 3 with Cv = 50 (USgpm/psi). Then only one flowrate is possible for the crossing between the two curves and one increases speed will give increased flowrate. When the flowrate has passed the point which gives the maximum head, one can reopen the choke and continue as normal.

4.7 System Protection

To maximize the lifetime of this first full-scale subsea processing installation, an important objective has been to outline operational procedures and control system logic's that gives good equipment and environmental protection and minimum wear, while maintaining simple operation for maximum availability.

An example of this compromise is that the number of valves that are closed automatically is kept to a minimum. This reduces valve wear and it avoids restart problems, although better protection against hydrates could have been achieved with more valves closed.

Base case for depressurization of Troll Pilot is made through the flowline to Troll C.
If one gets a situation with low pressure in the flowline and high pressure in the Troll Pilot separator, the flowline must be pressurized before the isolation valves between Troll Pilot and the flowline is opened. The reason is to avoid high pressure differentials which may damage the piping due to high velocities, impact forces and possible low temperature due to expansion.
Depressurization through the methanol injection system is not allowed due to risk of hydrate formation in the service line and unwanted amounts of hydrocarbons to topside.

The separator water outlet valve should not be operated frequently to avoid wear because it is located on the manifold module which is not intended to be retrieved. Instead the valve on the suction side of the pump and the bypass valve around the pump is closed on "low low" water level. These valves are located on the pump module which is defined as relatively easy retrievable. The reason for not just tripping the pump without closing valves is mainly to avoid forced rotation of the pump.

If the WI pump trips, the valve on the suction side of the pump is closed avoid forced rotation of the pump in both forward and reversed direction. The bypass valve is also closed to avoid low separator level.

Additional reasons for closing these valves are to:
- Reduce the risk of hydrate formation in the separator during shutdown, by ensuring that the liquid level does not drop below a certain limit to maintain a minimum thermal mass, and in the water injection system by avoiding oil and gas into the system.
- Prevent gas flow to the WI pump, to avoid damage to the pump.
- Protect the reservoir against injection of pure oil or water with high oil content.

The most important situations for which the pump is protected with automatic trips are:

- When any valve on the suction side or discharge side are closed, to avoid running the pump with closed suction or discharge.
- When the flowrate through the pump is below the minflow limit, to avoid heavy vibrations and high temperatures and to protect against discharge pressures above design.
- At high discharge pressure to avoid pressures above design.
- At low barrier fluid pressure margin to the pump suction pressure, to avoid water leakage from pump to motor.

5. SUMMARY AND CONCLUSIONS

Oil Production from the Troll field poses significant operational challenges due to the thin oil layers; such as a large span of water rates, gas to liquid ratios and subsea separator pressures. Hence, if success of Troll Pilot is demonstrated, the creditability of such systems should be high enough that other fields may also utilize this technology.

With the high expected water production to the Troll C platform, the Troll Pilot will make a significant contribution towards the reduction of produced water towards topside. Further, with the installation of Troll Pilot, the flexibility to handle an unforeseen production increase will be improved.

The full benefits from using subsea separation and reinjection technology will be unleashed the day a field is developed to fully rely on the performance of subsea separator stations. All depending on field characteristics and on existing nearby installations and infrastructure, this technology can contribute to:

- A significant reduction in the amount of produced water released to sea.
- A more compact topside process.
- Aid the tie in of smaller fields over longer distances to existing infrastructure.
- Reduced use of chemicals such as demulsifiers and possibly hydrate inhibitors.
- A reduced topside power consumption can be envisaged if an easy to maintain subsea separator station is located in close vicinity to or underneath a platform.
- Increased oil production because the wellhead pressure can be reduced.
- Less risk of "drowned" wells and enable/implify start-up (kickoff) of drowned wells because the wellhead pressure can be reduced.
- The flexibility to handle unforeseen production increases will be improved.
- With thin oil layers as on Troll, the horizontal well sections can be drilled deeper/closer to the WOC (water oil contact) and further away from the gas column. This will reduce the gas production and potentially increase the oil production and recovery.

For this first pilot installation, process design and operation aspects has been emphasized to reduce the risk of stopped or reduced oil production to Troll C. System operation has therefore been given a thorough investigation through the entire design phase.
The process design strategy has been to minimize utilization of new technology, since major developments was nevertheless unavoidable (high voltage wet mateable connector, subsea level detector, etc). This has resulted in a relatively simple process system with a large horizontal separator with the simplest possible internal configuration.

The project milestones is that engineering is completed, fabrication has been ongoing in parallel with the engineering phase, system testing will start in November 98 to be completed 1/7 99 and start-up on the field is planned to be late 99.

Further developments of the HYDROSEP™ system for downhole oil/water separation

I C SMYTH
Baker Hughes Process Systems
B FAY
Centrilift, UK

SYNOPSIS

Downhole separation is becoming established as an effective means of improving the economics and oil recovery of high water cut wells by substantially reducing the volumes of water brought to the surface. Applications, so far, have been land based but the extension of the technology to lower water cut, higher flowing wells, more typical of mature offshore fields, is of considerable interest to operators. A prototype installation using two stages of separation is described and its anticipated response in an imminent surface based trial discussed. The importance of process monitoring and control is also emphasised.

1. INTRODUCTION

1.1 The advantages of downhole separation

Increasing water cuts in oil wells are an economic and environmental challenge to the operator. Production rates fall as the well-bore fluid gradient and friction losses within the pipework increase. Ultimately, the capacity of production separators and water treatment equipment can be exceeded, necessitating choking back of the well which further reduces the flow of oil to the surface. Increasing surface processing facilities may not be an option due to space and weight limitations offshore or straight forward cost constraints for a well that is likely to be nearing the end of its life. In addition, surface disposal of produced water is a major environmental issue controlled by ever tightening legislation and for many onshore production sites it is banned altogether.

By separating oil and water downhole with re-injection of the water into a local disposal zone the efficiency of the production operation can be raised substantially. Higher oil production rates resulting from the removal of the constraining effects of the water can be further enhanced if the separated water can be used for reservoir pressure maintenance. Lower operating costs can also be anticipated, reflecting the energy savings in bringing much smaller fluid volumes to the surface (where artificial lift has been used) as well as reduced chemical usage (especially corrosion and scale inhibitors). The economics of more oil produced per unit cost will also improve the level of recovery from the reservoir (1).

Compared to the surface, downhole well-bore separation also benefits from higher fluid temperatures (which particularly reduces their viscosity), a lower oil gravity (because of dissolved gases) and a minimal degree of droplet shearing (in the absence of upstream valves). Not only is the separation easier, therefore, but acceptable residual oil levels in the separated water tend to be higher for injection below ground than for discharge above, allowing the downhole separation system to be more compact than its surface equivalent.

1.2 Downhole separation systems

In general terms, a downhole oil/water separation (DOWS) system must provide means for separating the oil and water, energy to dispose of the water and drive the oil to the surface and a way of isolating the production zone from the injection zone. Currently, the only proven oil/water separator for the downhole environment is the hydrocyclone. This reflects not only that the technology is well established for the surface treatment of produced water but, more particularly, its compact and robust nature. Pressure to drive the fluids through the hydrocyclone, produce the oil and dispose of the water may come from the reservoir itself but more commonly a well-bore pump is required. Whilst both beam and progressive cavity pumps have been successfully utilised in some completions, the most flexible and widely used option is the Electric Submersible Pump (ESP), particularly for high flow applications. Isolation within the well-bore is usually provided by a retrievable packer which is deployed independently.

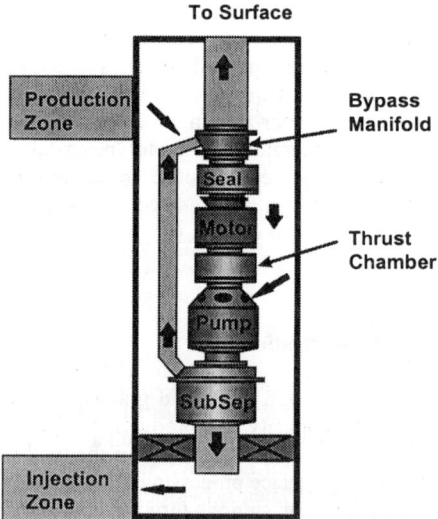

Fig.1 Single pump/single separation stage DOWS system (injection below production zone)

A simple system is illustrated in Fig.1. The production fluids are drawn into the pump which drives them through the separation module (or SubSep™). This comprises one or more hydrocyclone units, depending on flow requirements, hydraulically connected in parallel as shown in Fig.2. The pressurised fluids enter the hydrocyclones through their tangential inlets, forcing the flow to spin. The resulting centrifugal forces cause the separation of the oil/water mixture. Residual pressures lift the oil concentrate stream (discharged from the overflow) to the surface and force the water stream (discharged from the underflow) into an injection zone which is isolated from the production zone by a packer that forms part of the completion.

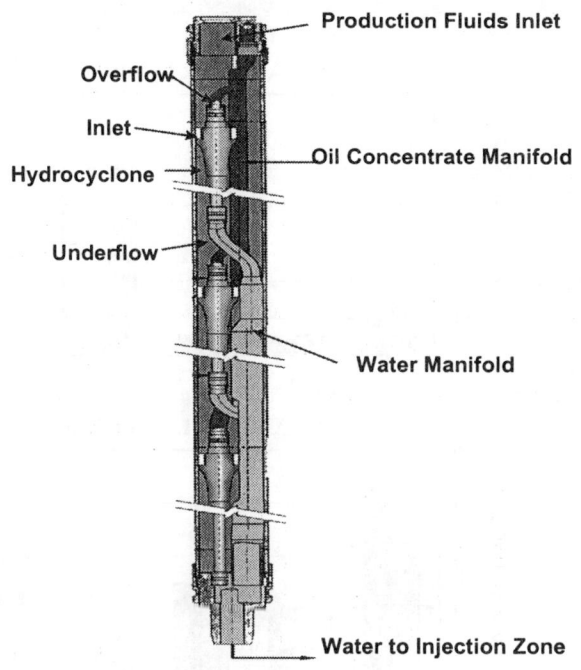

Fig.2 SubSep assembly with three hydrocyclones

Depending on reservoir conditions and the location of producing/injection zones, systems may also re-inject above the producing formation, have additional pumps and different completion formats (2) but all installations to date have used only a single stage of separation and been land based. These will be briefly discussed, however, it is the development of a high capacity, two stage hydrocyclone/two pump assembly suited to offshore applications which is the focus of the paper. This will be illustrated with particular reference to the HydroSep™ system, which is the generic terminology for a DOWS installation made up from Centrilift ESPs and Vortoil (Baker Hughes Process Systems) hydrocyclones.

2. EXISTING ONSHORE SYSTEMS

More than 40 DOWS systems have been installed to date, predominantly in relatively low flowing wells (60 - 650 m³/d) with high water cuts (90% or more). Systems tend to be set up at a fixed operating condition as it is not usually cost effective to have an active monitoring/control capability in such wells.

The potential impact of DOWS on oil and water production is summarised in Fig.3 for a number of a recent HydroSep™ installations. Whilst modest rises in oil production are evident, water flows to the surface have typically fallen by an order of magnitude. In terms of composition, the fluids coming up to the surface have changed from 2 - 6% oil to 18 - 33% oil. The quality of the disposed water is not normally measured but where it has been sampled, oil levels of a few hundreds of ppm have been found.

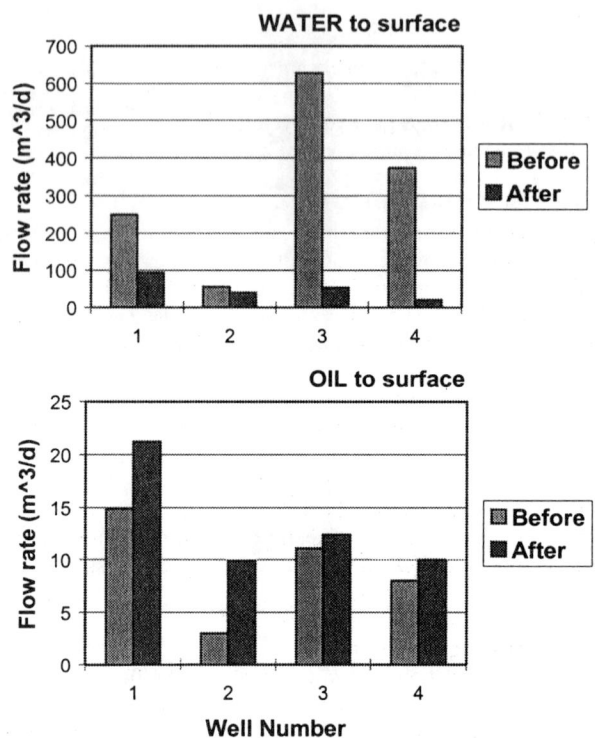

Fig.3 Effect of HydroSep installation on oil and water production

Since the first trials in Canada in 1995, a progressive improvement in system reliability has been observed, and run times of more than 500 days have been achieved. Where failures have occurred they have principally been due to problems with the ESPs and the injectivity of the disposal zone, although this has often been linked to poor initial characterisation of the injection formation (3).

3. POTENTIAL OFFSHORE SYSTEMS

3.1 Differences from onshore
In the offshore environment the costs associated with producing and treating water are even higher than onshore so the potential benefits of a downhole system are correspondingly

greater. In particular, savings in subsea processing/produced fluid transmission costs could be considerable for remote wells. Deployment risks will also be increased, however, as workover costs are higher and because wells tend to be more prolific the impact of deferred oil production is greater. This makes reliability a key issue and implies a increased level of monitoring would be desirable.

Offshore applications are also likely to have lower water cuts. This means that the separation process must get more efficient to achieve a particular discharge water quality as the oil content rises and whilst droplets may well become bigger, increased apparent viscosity of the mixture will work against the separation. A single hydrocyclone stage may function reasonably effectively up to 15% oil (or even 25% if the conditions are conducive to separation). Above this level (and probably up to the point of phase inversion so long as mixture viscosities do not exceed the 5 - 10 cP working limit of hydrocyclones) or if better water quality is needed, a two stage hydrocyclone system is recommended. This also provides the capability to achieve a more concentrated oil stream.

3.2 Offshore system trial

Such a two stage separation format has been adopted for trials of a HydroSep™ system designed to operate in North Sea wells. The unit has been built for deployment in a 245 mm (9 5/8") cased well to provide a nominal 3,200 m³/d (20,000 bpd) capacity over a range of water cuts between 50 and 90%. The system is being appraised as part of a JIP in a test well at a land based evaluation facility in Humble, Texas using a light crude (30-35 API) and a synthetic brine.

The primary objective of the study is to demonstrate that this more complex installation can function effectively in the well environment so as to increase confidence in running a field application. Importance is also attached to the ability of the system to respond to changing conditions downhole and to sustain both a clean water discharge (target <500 ppm oil) and a concentrated produced oil stream (target >90% oil). In addition to variation in influent water cut, the test programme intends to look at different injectivity and productivity indexes.

At the time of writing, the DOWS system is being commissioned and the tests are imminent.

3.3 System design and envisaged operation

The general layout of the installation as it might appear in the field is shown in Fig.4. In the tests, however, the well fluid feed is provided from the surface, the clean water disposal stream also returns to the surface and there is no downhole oil/water monitor. A simplified flow schematic incorporating some typical pressure/flow data generated by Centrilift's HydroSep™ modelling programme is given in Fig.5

Pumps - It can be seen that a separate ESP has been used to feed each separation stage. The primary pump (562 Series KC 20,000) generates the greater head, sufficient to drive the oil stream to the surface when water cuts are low, the secondary pump (562 Series KC 16,000) is optimised for lower throughputs and pressure rises. The flow in from the production perforations is directed past the common motor to aid cooling. It is the choice of a "push through" pumping arrangement (having the pump upstream of the separator rather than downstream in "pull through" mode) which makes it possible to run with a single variable speed motor for both pumps, simplifying control. As a result, produced fluid homogeneity is also maximised and any free gas is dispersed and tends to be put back into solution, both factors which benefit hydrocyclone operation. The draw back is that, depending on the pump geometry/hydraulic efficiency and the interfacial chemistry of the fluids, the shear experienced in passage through the pumps may break up the oil droplets to such an extent that hydrocyclone performance is affected. Evidence from surface centrifugally pumped deoiling hydrocyclone applications suggests this is usually not a significant problem, especially if the pump has limited pressure rise per stage and is working close to its BEP. The high oil concentrations involved in this particular application will also mean that coalescence becomes a important process between pump and separator, which further strengthens the argument for the push through format. Whilst the pull through operational mode avoids the shear problem, it is more prone to constraints on separation due to free gas and poor homogeneity of the inlet mixture and system flow control is inherently more difficult.

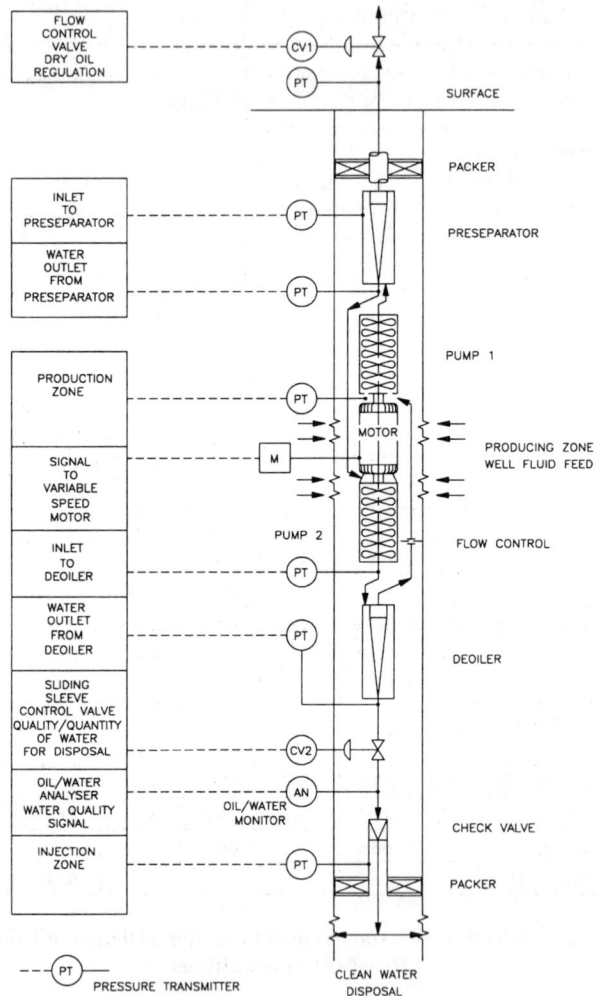

Fig.4 Hydrosep system with two stage separation

Separators - These comprise two SubSep™ units of 10 Vortoil high capacity G_M hydrocyclones (assembly diameter 178 mm ($7^5/_8$")). The first stage hydrocyclones are "preseparators", having a large overflow capacity suited to dealing with bulk oil. The second stage hydrocyclones are "deoilers", designed to cope with relatively low oil levels. The operating envelope for the units is 1200 - 4000 m³/d at pressure drops (inlet to water) of 1.4 - 15 bar. The lower limit is a function of separation efficiency and the upper limit a function of erosional velocity.

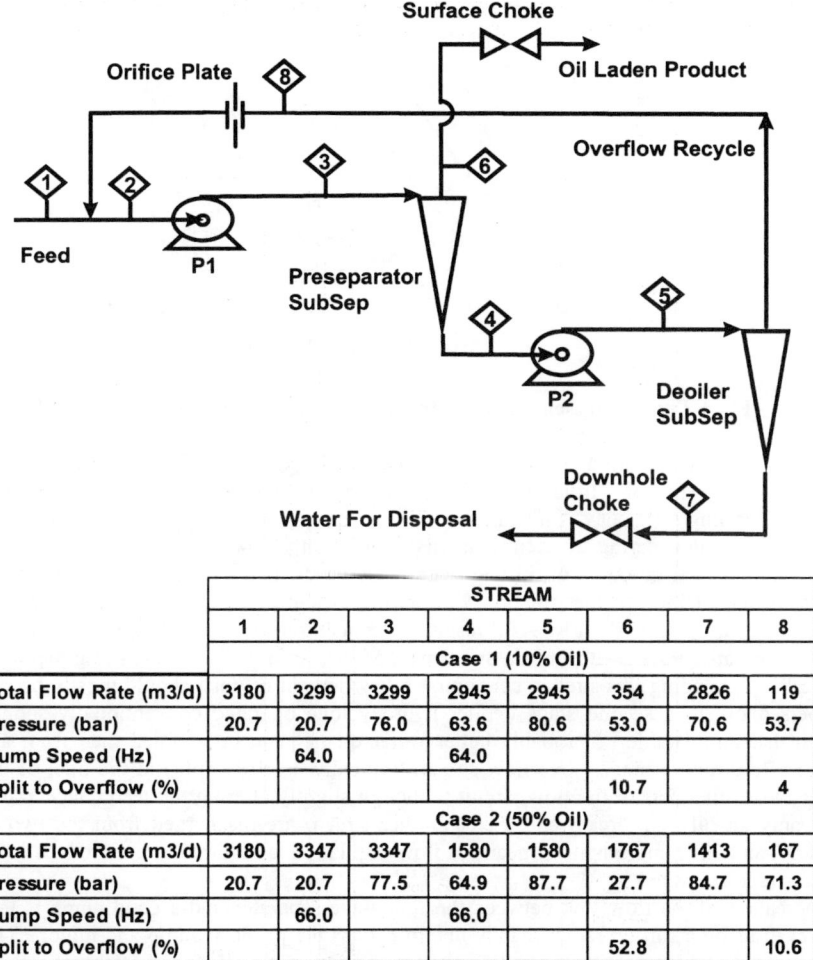

	STREAM							
	1	**2**	**3**	**4**	**5**	**6**	**7**	**8**
	Case 1 (10% Oil)							
Total Flow Rate (m3/d)	3180	3299	3299	2945	2945	354	2826	119
Pressure (bar)	20.7	20.7	76.0	63.6	80.6	53.0	70.6	53.7
Pump Speed (Hz)		64.0		64.0				
Split to Overflow (%)						10.7		4
	Case 2 (50% Oil)							
Total Flow Rate (m3/d)	3180	3347	3347	1580	1580	1767	1413	167
Pressure (bar)	20.7	20.7	77.5	64.9	87.7	27.7	84.7	71.3
Pump Speed (Hz)		66.0		66.0				
Split to Overflow (%)						52.8		10.6

Fig.5 Schematic of Hydrosep (two stage separation) and anticipated operation at typical Humble test conditions

The operational philosophy behind having two stages of separation to achieve one high quality oil and one high quality water stream is illustrated by reference to Fig.6. This shows how the quality of the discharge streams from a single preseparator hydrocyclone depends on the flow split (fraction of the inlet flow leaving by the overflow) relative to the feed oil concentration. The curves are of a generalised form and have been generated from a combination of field and laboratory experience for oil/water systems that will readily separate. A high dewatering efficiency reflects an oil rich overflow stream, a high deoiling efficiency reflects an oil free underflow stream[1].

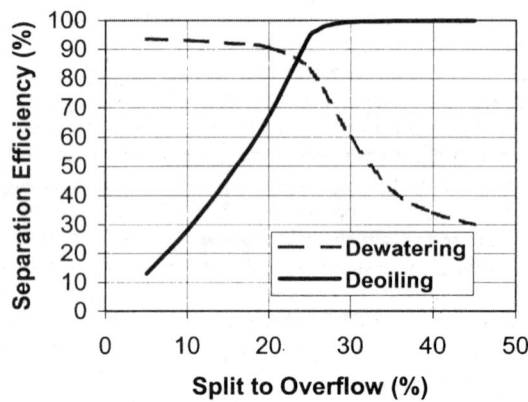

Fig.6 Typical preseparator hydrocyclone performance for a 25% oil feed

Conventional operation, to achieve a water discharge with minimum oil, would mean running with a split of a little over 30% for the 25% oil feed condition being considered so that the deoiling efficiency is maximised. Whilst the water stream may then contain around 500 ppm oil, dewatering efficiency at this split is only modest with the overflow stream containing similar amounts of oil and water. However, if the operating split is dropped to close to that of the oil inlet fraction, dewatering efficiencies climb sharply whilst deoiling efficiencies drop only slightly. As a result, oil concentrations in the overflow stream can be pushed towards 90% with only a small part of the separated oil core being "spilt" to the underflow. Sending this stream, containing a few percent of what should mostly be coarsely dispersed oil, to a second stage deoiler hydrocyclone operated conventionally should also mean that comparable or probably better water quality can be attained than from a single stage. Returning a modest overflow stream (typically 4 - 10% split) to the inlet of the first stage closes that part of the flow circuit (although slightly altering the feed concentration) so that only an oil rich stream and a clean water stream are discharged from the two stages. Clearly, an active and sensitive means of flow regulation is required to get the best out of the system (see below).

A feature of the flow link between the first stage separator and second pump is that it is achieved using a shroud rather than a tube to bypass the primary pump and motor. This is a more robust arrangement that is better suited to the deviated wells found offshore.

Monitoring and control – Downhole conditions (reservoir pressure, water cut, injectivity, etc) will alter with time and an effective DOWS system needs to be responsive to such changes. This implies an ability to measure process conditions so that operation can be

---

[1] Dewatering efficiency = 1- $((1 - C_o)/(1 - C_i))$; deoiling efficiency = 1 - (C_u/C_i);
where C = oil concentration in hydrocyclone overflow (o) ,underflow (u) and inlet (i)

adjusted to maintain performance. Monitoring is particularly critical during start up as risk of failure is highest then. The use and extent of monitoring equipment will be an economic balance of the cost/risk factors associated with the well.

The most basic level of instrumentation is pressure measurement and six static pressure sensors are included in the downhole assembly (Fig.4) covering the production and injection zones and either side of the separation stages (pressure at the preseparator overflow is inferred from a surface measurement). This allows reservoir conditions and pump operation to be directly monitored and flow through and flow split from the separation stages to be determined (from pressure drops and pressure drop ratios respectively).

A means of monitoring oil/water concentrations is also desirable. In the Humble tests, complete characterisation of the flows to and from the installation is possible. The make up of the inflow is known as it is set up at the surface, the water stream is also returned there to facilitate analysis and to allow different injectivity conditions to be simulated by applying back pressure to the flow, whilst the produced oil stream passes through a BS&W meter. In the field, the production stream will be the most accessible and easiest to instrument and although this would provide some indication of separator operation downhole, it would be preferable to monitor well inflow oil concentrations and residual oil levels in the injected water for best control of the system. Of these, it is probably most critical to measure the quality of the disposal stream, as oil carryover into the injection zone is more of a risk to the effective operation of the DOWS system than some additional water being produced to the surface. At present this can only be practically achieved by bringing a sampling capillary to the surface but ideally a downhole oily water monitor is required. Instruments based on light scattering are marketed for surface use but have proved susceptible to interference from gas and solids. Fluorescence based techniques are showing promise, though, and a device using a free falling sample stream which does not contact any optical components has recently been approved for continuous monitoring of discharges from a Norwegian platform. Such a sensing cell would not be practical for downhole use but a cell capable of high pressure operation has been developed at Heriot-Watt University as part of a photo-acoustic device for surface based oily water concentration measurement (4). Offshore field trials are planned for later this year, downhole testing should be the next step in the development.

Flexible flow and pressure control is achieved by a combination of pump speed variation and the use of a hydraulically activated downhole choke on the disposal stream and/or a surface choke on the produced oil stream. For example, the conditions shown in Fig.5 represent the extremes of the oil concentrations to be evaluated in the Humble tests. In moving from the high to low oil condition whilst maintaining overall throughput at 3180 m^3/d, injection pressures of 70 bar and oil discharge pressures sufficient to produce to the surface (i.e. nominally >21 bar), the downhole choke would be opened up and the surface choke closed down to reduce the flow split to the overflow of the preseparator stage in line with the feed oil concentration change. The higher flows to the deoiler stage means that its overflow pressure falls so the recycle rate, which is regulated with a fixed orifice plate, also drops. A slight reduction in pump speeds can also be accommodated in this process, although if the 4% recycle flow was considered too low, a higher motor speed would allow it to be increased.

In the trials, the adjustment to changing process conditions will be carried out manually but if the installation were fully instrumented the response could be automated allowing remote operation in the field. Such an intelligent completion would also benefit from incorporating condition monitoring equipment (e.g. temperature and vibration sensors) to assess the health of the motor/seal/pump assembly so that any interventions could be planned (5). It is likely that offshore and deep water applications would require completions of this kind.

Gas and solids - Operating at lower water cuts also means an increased likelihood of gas being co-produced and gas levels of 10% by volume at the influent pressure conditions are to be considered as part of the trial. This is unlikely to affect either pump operation or preseparator hydrocyclone performance to any degree, especially with the push through arrangement of the equipment.

Whilst solids are not part of the current test programme, if sand is produced in any significant quantities there will be implications for the long term injectivity of the disposal formation and possible erosion problems for the hydrocyclones. Although fundamentally dependent on the nature of the producing zone formation, higher flowing wells also tend to produce more sand. These solids are likely to be water wetted and all but the very finest material will tend to go with the water stream through the hydrocyclones and end up in the injection zone. Nominal upper limits of 100 mg/l and 50μm are appropriate to avoid erosion problems and oil/water separation will be unaffected at these concentrations.

A number of approaches can be adopted to deal with solids, either individually or in combination. At the simplest level, gravel pack, resin treatments and screens could be employed in the well to keep the coarser particulates out of production. Sand could also be removed in a dedicated hydrocyclone as part of the separator train with the solids being re-entrained in the oil stream to be dealt with in a surface facility. Alternatively, enough pumping power could be provided to dispose of the water above the injection zone fracture gradient.

4. IMPLICATIONS FOR RESERVOIR MANAGEMENT

Beyond the direct benefits that installing a DOWS system in a well can generate, like reduced power costs and the freeing up of surface equipment, there is often considerable potential to use the separated water as a flooding mechanism reducing the need for additional water injection boreholes, turning a producer into a producer/injector. With appropriate completion design, this could be cross-flooding between adjacent producing wells or horizontal-flooding in a multilateral (6). Other downhole production methods which might be adopted include the inverse coning technique. This can be applied in highly permeable reservoirs with strong bottom water drive where coning of the water into the oil perforations is prevented by separate production of water from further down the same wellbore. This water can then be pumped into a lower, isolated disposal zone (2). As limited produced fluids are brought to the surface, an ability to monitor and control injection volumes and water quality would generally benefit these type of applications

5. CONCLUSIONS

Downhole oil/water separation has been found to be an effective tool in revitalising the productivity of high water cut wells, especially where lifting costs are high and surface treatment facilities constrained.

The potential for taking the technology into high flow (3,200 m³/d), low water cut (down to 50%) wells, more typical of offshore fields, is currently being evaluated using a prototype system with two stages of separation. This should be capable of providing both a high quality water stream for downhole disposal and a concentrated oil stream (perhaps 90% oil) returning to the surface. Monitoring and control systems are also being investigated as part of the present development so that the installation can be made responsive to process changes. Progress in these areas would also benefit the exploitation of the technology as a reservoir management tool.

The integration of intelligent systems with downhole separation systems will be necessary for the effective use of the technology in offshore applications. Reliability will remain a key issue but the potential savings in subsea processing/produced fluid transmission costs could be considerable for remote wells.

ACKNOWLEDGEMENTS

The authors would like to thank the management of Baker Hughes Process Systems and Centrilift for permission to publish this paper.

REFERENCES

1) Peachey, B.R. and Mathews, C.M., "Downhole oil/water separator development", J of Canadian Petroleum Technology, Vol.33, No.7, Sept 1994.

2) Loginov, A. and Shaw, C., "Completion design for downhole water and oil separation and invert coning", SPE Paper No.38829, SPE Annual Technical Conf, 1997.

3) Foulser, B., "Water management - downhole oil/water separation", SPE Review, June 1998.

4) MacKenzie, H.A., Freeborn, S. and Hannigan, J., "A pulsed photoacoustic instrument for the detection of crude oil concentrations in produced water", Water Management Offshore Technical Briefing, May 1998, Oslo, IBC (EJ129).

5) Tubel, P. and Herbert, R.P., "Intelligent system for monitoring and control of downhole oil water separation applications", SPE Paper No.49186, SPE Annual Technical Conf, Sept 1998, New Orleans.

6) Chrusch, L.J., "Downhole oil and water separation - potential of a new technology", Proceeds. 25th Silver Anniversary Convention of the Indonesian Petroleum Association, Oct 1996.

CMPT – brokering solutions for the upstream oil and gas industry

J SHACKLETON
CMPT, Aberdeen, UK

The Centre for Marine and Petroleum Technology (CMPT) is a not-for-profit company providing opportunities for research and development organisations to access the latest advances in science, engineering and technology (SET).

CMPT does not undertake RTD & D in-house. Instead, it operates the CMPT Virtual Trading Floor™, 'dealing' in information, knowledge and know-how relating to both SET and to the supply chain process associated with RTD & D and innovation. CMPT's Virtual Trading Floor™ and its staff of researchers and brokers aim to connect:

- what industry needs; with
- what the research community and technology delivery companies can provide; with
- what the public sector can do to assist.

CMPT provides the following services:

- SET-Screen™. Assisting companies in the evaluation / preparation of project proposals.
- SET-Finder™. Identifying capability and expertise in science, engineering and technology for its Member companies (the major oil and gas exploration and production companies).
- SET-Finance™. Identifying the most appropriate funding sources for oil and gas projects.
- SET-Monitor™. Highlighting and reporting on oil and gas operators' technology needs.
- SET-Sourcing™. Finding complementary partners for public sector funded projects.

In addition, CMPT is actively involved in the set-up and launch of RTD projects for joint industry funding with particular emphasis on specific themes identified as being of particular importance to CMPT's Member companies. Current themes include:

- **Total Elimination of Surface Facilities for Deepwater and Very Small Marginal Field Production**

Totally eliminating the need for surface facilities would represent a major step forward for the industry. It would have a major impact on costs, on safety and on environmental impact.

Many individual technologies which form potential building blocks to achieve Zero Surface Facilities Capability (ZSFC) are already either available or under development. What appears to be missing is a comprehensive appraisal of whether these building blocks are the right ones, can they be integrated into an effective system, and where are the gaps?

Topics to be addressed, including some issues which are already receiving attention, include: power generation and distribution; electric actuators; sub-sea separators; raw water injection, sand disposal, down-hole processing; down-hole compression; down-hole pumping; smart completions, new sensors; more "intelligent" control systems; high-bit rate acoustic telemetry and intervention techniques.

- **Super-Smart Production Systems**

Smart production systems, which permit the regulation of flow from different parts of the reservoir, are already in existence. "Super Smart Production Systems" will not only measure flow but analyse the chemistry of the produced fluids and then automatically operate down-hole chokes to close off or open up producing intervals and modify the characteristics of the subsequent process train in real-time. A number of new or enhanced technologies may be required including:

- new sensors to measure fluid chemistry and other parameters at the interface between reservoir rock and the production tubing;
- data transmission, analysis and systems to modify down-hole valve positions, inject treatment chemicals and regulate other parts of the process system characteristics etc;
- a process system whose characteristics can be modified to suit a wide range of production condition over time.

- **Environmental Impact Reduction**

Probably one of the most critical factors facing the industry today, environmental impact reduction and the related SET issues are concerned with both regional exploration and production activity and the global 'greenhouse' effect. CMPT will enter into close consultation with Member companies and their representative organisations in framing the details of its RTD&D strategy.

- **Deeper/Higher Resolution Imaging of Rocks and Fluids in the Sub-surface**

The prize for significant improvements in subsurface imaging of rocks and fluids includes fewer dry holes, enhanced well planning and improved recovery.

Options here include further improvements in seismic technology or the development of an alternative. For example, recent experiments making small areas of the ionosphere produce artificial and localised "aurora borealis" have led to the discovery that charged particles penetrate the Earth and can be used to detect subsurface features. Does this hold potential for the oil/gas industry?

- **Deepwater Drilling Cost Reduction.**

The high costs of deepwater drilling and the subsequent effect on exploration has prompted CMPT to evaluate the process of how offshore drilling operations are undertaken. Results of recent studies have demonstrated that costs have risen significantly over the past five years, causing, amongst other things, a fifteen percent drop in Gulf of Mexico activity and an increase in spending in the upstream oil and gas industry of around 4 billion dollars. The challenge, therefore, faced by CMPT and the upstream oil and gas industry is to reduce the cost differential between shallow and deepwater drilling, whilst at the same time reducing overall drilling costs.

In light of these investigations, CMPT has launched a Joint Industry Project (JIP) which proposes the concept of Drilling Independent of Depth – DIODe. The basis of the project is to adapt coiled tubing technology for seabed well construction and intervention operations. It has been evolved not just to save costs, but also to initiate new thinking about how the whole process of offshore drilling operations are undertaken. DIODe is perceived as a mobile unit capable of being easily moved from one seabed location to another with minimal surface facilities, and, ultimately, no surface facilities at all.

CMPT Virtual Trading Floor™

Details of all CMPT's activities and assistance in finding solutions to problems can be obtained by contacting a broker on CMPT's Virtual Trading Floor™. The Trading Floor provides access to:

- state of the art technologies currently being used in the oil and gas industry;
- new technology solutions resulting from R&D;
- state of the art technologies and R&D in other industry sectors which have potential applications in the oil and gas industry;
- research and development capability in universities, research organisations, SMEs, and contractors;
- research and development funding sources including operators, government, contractors and investors.

The CMPT Virtual Trading Floor™ is available via: Direct Access: +44 (0)870 608 3434. Online Access: www.cmpt.com. E-mail: broker@cmpt.com

Authors' Index